S250 Science in Context
Science: Level 2

The Open University

Introduction to the course
and
TOPIC 1 BSE/vCJD

Prepared for the Course Team by Richard Holliman and Pat Murphy

This publication forms part of the Open University course S250 *Science in Context*. Details of this and other Open University courses can be obtained from the Student Registration and Enquiry Service, The Open University, PO Box 197, Milton Keynes, MK7 6BJ, United Kingdom: tel. +44 (0)870 333 4340, email general-enquiries@open.ac.uk

Alternatively, you may visit the Open University website at http://www.open.ac.uk where you can learn more about the wide range of courses and packs offered at all levels by The Open University.

To purchase a selection of Open University course materials visit http://www.ouw.co.uk, or contact Open University Worldwide, Michael Young Building, Walton Hall, Milton Keynes MK7 6AA, United Kingdom for a brochure. tel. +44 (0)1908 858785; fax +44 (0)1908 858787; email ouwenq@open.ac.uk

The Open University
Walton Hall, Milton Keynes
MK7 6AA

First published 2006.

Edited and designed by The Open University.

Typeset by The Open University.

Printed and bound in the United Kingdom by CPI, Glasgow.

ISBN 0 7492 1433 3

1.1

The S250 Course Team

Andrew J. Ball (*Author, Topic 2*)

John Baxter (*Author, Topic 6*)

Gerry Bearman (*Editor*)

Steve Best (*Media Developer*)

Kate Bradshaw (*Multimedia Producer*)

Audrey Brown (*Associate Lecturer and Critical Reader*)

Mike Bullivant (*Course Manager*)

James Davies (*Media Project Manager*)

Steve Drury (*Author, Topic 3*)

Lydia Eaton (*Media Assistant*)

Chris Edwards (*Course Manager*)

Mike Gillman (*Author, Topic 4*)

Debbie Gingell (*Course Assistant*)

Sara Hack (*Media Developer*)

Sarah Hofton (*Media Developer*)

Martin Keeling (*Media Assistant*)

Richard Holliman (*Course Themes and Author, Topic 1*)

Jason Jarratt (*Media Developer*)

Simon P. Kelley (*Author, Topic 2*)

Nigel Mason (*Topic 7*)

Margaret McManus (*Media Assistant*)

Elaine McPherson (*Course Manager*)

Pat Murphy (*Course Team Chair and Author, Topic 1*)

Ian Nuttall (*Indexer*)

Judith Pickering (*Media Project Manager*)

William Rawes (*Media Developer*)

Shelagh Ross (*Author, Topic 7*)

Sam Smidt (*Author, Topic 7*)

Valda Stevens (*Learning Outcomes and Assessment*)

Margaret Swithenby (*Media Developer*)

Jeff Thomas (*Author, Topics 6 and 7*)

Pamela Wardell (*Media Developer*)

Kiki Warr (*Author, Topic 5*)

The Course Team would like to thank the following for their particular contributions:
Benny Peiser (*Liverpool John Moores University; Author, Topic 2*), David Bard
(*Associate Lecturer; Author, Topic 6*) and Barbara Brockbank (*Associate Lecturer;
Author, Topic 6 and Critical Reader*).

Dr Jon Turney (*University College London and Imperial College London*) was
External Assessor for the course. The External Assessors for individual topics were:
Professor John Mann (*Queen's University, Belfast*); Professor John McArthur
(*University College London*); Dr Richard Reece (*University of Manchester*); Dr Rosalind
M. Ridley (*University of Cambridge*); Dr Duncan Steel (*Macquarie University, Australia*);
Dr David Viner (*University of East Anglia*) and Professor Mark Welland FRS
(*University of Cambridge*).

Contents

Introduction to the course

Welcome to S250 *Science in Context*. This course deals with a range of scientific disciplines, facilitating comparison between a series of scientific topics that have been chosen to represent the importance of scientific information in contemporary society. It is useful to acknowledge from the outset that you may find the approach in *Science in Context* is slightly different from that taken in other science courses you have studied. This is for two main reasons:

- First, you will study a series of seven scientific topics, focusing on specific aspects to illustrate important issues of scientific interest. This approach to science learning may contrast with your previous experience of studying disciplines such as biology, chemistry, Earth sciences and physics as separate subjects. A key aim for the course is therefore to increase your overall understanding of contemporary scientific knowledge and the ways in which science is conducted through comparison of what is currently known about these seven topics.

- Secondly, you will study how these scientific topics interrelate with four themes: communication, risk, ethical issues and decision making. Although this list of themes is far from exhaustive – we could have chosen several others – consideration of them can provide interesting insights into the ways in which scientists and scientific topics interact with the wider world. A further aim of the course is therefore to develop your understanding of how the topics have been influenced by these themes, facilitating reflection on how scientific knowledge shapes society and how society influences the progress of science.

Overall, the course aims to develop your knowledge of science and your scientific skills, whilst providing an introduction to some of the contextual factors that influence how scientific topics emerge as issues of academic study and public discussion. By the end of the course, you will have acquired skills in comparing and contrasting the seven topics, developing knowledge of how the science and themes interrelate to influence our understanding of these issues. These skills will not only be of use in your future studies, they should also be valuable if you choose to investigate scientific issues of interest to you in your everyday life.

Scientific knowledge can be very visible in contemporary society, not least when scientific and science-related issues are subjected to public discussion. As a result, scientific knowledge circulates not only within academic disciplines (e.g. in laboratory discussions, research papers and presentations at academic conferences), but also within wider society (e.g. on television and radio, and in newspapers and magazines, popular science books, museum exhibits, political debates, public meetings and Internet chat rooms). Examples of scientific topics that have generated public discussion in recent years include:

1   The emergence of the cattle disease BSE (bovine spongiform encephalopathy) and its impact on the beef and dairy industry, and the detection of the fatal human brain disease, vCJD (variant Creutzfeldt–Jakob disease).

2   The potential for asteroids or comets (known as near-Earth objects) to collide with the Earth, with potentially catastrophic consequences for human civilization.

3   The public health crisis in Bangladesh and West Bengal where millions of people have been drinking water containing dangerously high levels of the poisonous chemical element, arsenic.

4   The medicinal use of plants, such as St John's wort, which has been used in the treatment of headaches and mild to moderate forms of depression.

5   The debate over whether we are living through a period of significant global climate change, whether this is human-induced, and the role of humans in managing the environment.

6   The prospects for developing novel genetically manipulated materials for use in agriculture and in medicine.

7   Developments in nanotechnology and the prospects for using these techniques in the production of novel medicines and materials.

You should now complete the following activity.

Activity

Allow 10 minutes

Take a few moments to consider the seven examples listed above. Had you heard of all seven examples prior to starting the course? Choose one example that is of particular interest to you to investigate further. Once you have chosen your example, try to answer the following questions, making brief notes as you proceed. Where did you first hear about your chosen example? Did you actively seek further information about this topic? If so, where did you look for this information? What are the key facts and issues that you know about this scientific topic?

Comments on activity

Each of the seven examples has generated a considerable amount of science communication in recent years and this makes them high-profile examples, both within the scientific community and wider society. As such, it is possible that you may have heard of all of them. Having said this, an enormous number of announcements of scientific progress are made each year. Given the speed of this scientific progress, it would be impossible to keep track of all the many and varied scientific developments, even in relation to just these seven topics. As consumers of scientific information you make selections, and these will often be based on what you find interesting and relevant to your everyday life. For example, it is possible that you have a particular interest in one of the topics, perhaps because you stopped eating beef following the BSE/vCJD episode or you have friends or relatives who are affected by contaminated water in Bangladesh and West Bengal. At the same time, you might regularly recycle your newspapers and glass bottles, or compost your food waste, having read articles reporting scientific research into global climate change. By the same token, there may be issues that you have not heard of, or that you have only heard of in passing. The newly emerging area of nanotechnology might be just such an issue, in part because the study of this area is still in its infancy. Overall then, it would be fair to assume that you have a number of interests and reasons for studying this course, and varying levels of prior knowledge and experience of the topics. This prior knowledge and experience is also likely to have influenced your attitudes and behaviour; for instance, you might have chosen to avoid genetically manipulated foods because of a concern about the long-term safety of those foods for human health.

It is likely that the first time you heard of these issues would have been through news media coverage, for instance through watching a television news bulletin, listening to a radio broadcast or reading a newspaper article. Indeed, research suggests that news media continue to be an important source of information after we have left school, particularly about newly published science. In this respect, it is common for people to regularly access sources of information about science that they consider to be credible and trustworthy. Perhaps you subscribe to a popular science magazine or have an online news outlet as the home page on your personal computer. Maybe there are other sources of information that you may have included in your answers to this activity, including your friends and family, Open University courses, films, the Internet, your doctor or local health practitioner, even a local farmer or butcher in the case of BSE/vCJD. It is therefore perfectly possible to find out about scientific topics from everyday sources that do not involve newspapers, television or radio. Rarely, however, will non-scientists access the academic research papers written by scientists. Instead, non-scientists tend to rely on third parties, such as journalists, who are trained to mediate information for public consumption, or on friends, family and everyday acquaintances who are considered to be reliable and knowledgeable.

It is more difficult to judge whether you will have sought further information about any of the examples. You may have been more interested in one example when compared to another, investigating a range of sources on this topic, including books and web-based information. You might even have chosen to study S250 based on your initial interest in one or more of these topics! We hope that this will be the case as these seven topics form the basis of the course. Of course, these are not the only scientific topics to have become the subject of public discussion in recent years and a selection had to be made. It is worth noting therefore that these topics were chosen by the Course Team, in part because they were of interest to the authors but also because they raise interesting aspects of the themes. What then can you expect to cover as you study S250?

The course topics

The following descriptions provide brief outlines of the seven topics. Compare these outlines with your notes from the previous activity to see how much knowledge and experience you already have about your chosen topic. Don't worry if your list looks a lot shorter than the paragraphs below. Indeed, we hope that this will be the case as all the information you need to know about each of the topics to complete the assessment will be taught as you progress through the course.

Topic 1 *BSE/vCJD*

Bovine spongiform encephalopathy (more commonly known as BSE or 'mad cow disease') and variant Creutzfeldt–Jakob disease (vCJD) were unknown prior to the mid-1980s. Since then, these diseases have been subjected to ongoing scientific studies and public debate involving, amongst others, scientists, politicians, vets, farmers and the public. What scientific knowledge is now known about BSE/vCJD and what are the continuing uncertainties? How did decision makers manage the risks associated with BSE/vCJD and communicate them to the public? What has been the impact of the BSE/vCJD episode on the relationship between science and the public? This topic will address these questions by outlining key events in the emergence of these diseases. You will also learn about the biology of BSE and vCJD, and the importance of epidemiological research in developing understanding of new illnesses and their distribution in a given population. The four themes of communication, risk, ethical issues and decision making will be addressed, for example through consideration of the importance of risk communication by politicians and scientists.

Topic 2 *Near-Earth objects and the impact hazard*

With the development of space flight, alongside other astronomical research, scientific understanding of the Solar System, and in particular near-Earth objects (NEOs), has increased immeasurably. But is the collision of the Earth with asteroids and comets really a danger to our world? If so, what sort of threat do NEOs pose and how should we prepare? How do we assess the risk of these extremely rare but potentially catastrophic events, and what might we do to mitigate the threat? In the NEOs topic, you will explore the science of astronomy to learn about asteroids and comets. You will find out how big they are, how many have been identified, where they come from, and how astronomers have learnt more about them. You will look at the record of impact craters on the Earth to find out what happened when objects hit in times past. You will see what effects a large impact can have beyond the hole in the ground, and consider some of the evidence for the effects of small and large impacts, including a massive impact 65 million years ago. Through discussion of the themes of communication, risk, ethical issues and decision making, you will also consider how society has reacted to the increasing scientific knowledge of NEOs and how this information is disseminated to the wider world.

Topic 3 *Water and well-being: arsenic in Bangladesh*

In the 1990s, drinking water supplies to tens of millions of people in Bangladesh were found to be contaminated with the poison, arsenic. Research following the emergence of arsenic-related illnesses revealed that the contamination was due entirely to natural causes. In this topic, you will learn that this contamination relates to the alluvial sediments from which groundwater is extracted, and how unusual chemical conditions in them release arsenic from a common mineral. The risks from water-borne arsenic are complex; understanding them demands not only geoscientific information, but also knowledge of the ways in which arsenic affects human metabolism once it is present above threatening levels, themselves a subject of debate. Communicating these risks to those affected is therefore an important issue that is addressed in this topic. This ongoing situation poses dilemmas for most of the local population and for regional, national and international decision makers as regards deciding how to mitigate the effects of chronic arsenic poisoning. This is further complicated because British Government scientists missed an opportunity to detect the problem before the effects of water-borne arsenic became obvious, raising important ethical issues for geoscientists. You will investigate these issues through consideration of a legal claim for compensation by affected Bangladeshi villagers against these scientists.

Topic 4 *Medicinal plants*

Throughout history, plants have been used by humans for a range of therapeutic purposes, for example as pain-killers, muscle relaxants, protection against malaria, dietary aids and contraceptives. In the medicinal plants topic, you will explore aspects of the discovery, use, properties and development of a variety of plants and their compounds. This topic reveals the role played by different types of communication, from the oral traditions of indigenous peoples, via herbals, to research papers in scientific journals and web pages on the Internet. In so doing, you will be introduced to the means by which knowledge has passed from a few people to a wider audience, including researchers and pharmaceutical companies, and the implications this has for intellectual property rights. The topic probes into the biochemical mechanisms underlying the beneficial and harmful effects of plants and considers the evolutionary processes that have generated these mechanisms. The wide range of effects of plant compounds on humans raises a series of ethical issues as well as questions of risk, which are discussed in relation to the requirements for regulation of medicines through decision making processes.

Topic 5 *Climate change*

Climate change has long been an issue of scientific study and public interest. It is now a major contemporary environmental issue attracting widespread scientific interest that produces evidence which, in turn, informs decision making at a number of different levels right through to international treaties such as the Kyoto Protocol. Why is climate change such a high-profile issue and why does it matter? In the climate change topic, you will be introduced to the main elements in the climate change story, acknowledging the early work completed by scientists studying levels of carbon dioxide in the atmosphere in the 1950s, and the interventions of environmental campaigners in the 1970s. You will learn about the science behind terms such as 'global warming', the 'greenhouse effect' and the 'global climate system', and what types of evidence inform scientific understanding of these concepts. You will consider what role human activity has played in generating the conditions for climate change and also how these issues are being managed, drawing on the themes of communication, risk, ethical issues and decision making.

Topic 6 *Genetic manipulation*

Genetic manipulation is a broad term that covers issues such as cloning, genetically modified (GM) crops and gene therapy. A greater awareness of the influence genes have on our lives, and those aspects of our behaviour and biology where genetic influences form one part of a more complex picture, raises new social challenges and ethical issues. This topic describes how genes are manipulated, producing new combinations within animals and plants that aim to modify what we generally understand as 'natural'. Are such innovations taking place so fast and so often that the opinions of 'ordinary people' are being left behind or overlooked? If so, what measures can be taken to ensure that the public are engaged in deliberations about genetic manipulation? Innovations in the use of genetics in agriculture and medicine are described as a prelude to outlining some of the ethical issues that such advances raise. How should knowledge gained in such areas be used to best effect? What risks are inherent in the adoption of these new techniques? What type of decision making will be required as genetic understanding impinges more on our daily lives and how can the associated benefits and threats best be communicated?

Topic 7 *Nanotechnology*

Developments in the science of nanotechnology are forecast to play an increasing role in our everyday lives in the 21st century. They may bring benefits in areas as wide as drug delivery, remediation of water contamination, and the development of stronger, lighter materials. As a result, the area is attracting major investment from governments and business. In the nanotechnology topic, you will first look at what it means to work at the nanoscale. You will then examine two areas in more detail: first, the nanoscience of materials and second, nanotechnology and living systems. You will look at the ways in which we might learn from nature to create bio-nanotechnological devices which could be integrated with living systems, whilst considering the possible new risks and ethical issues involved with the insertion of adapted biomaterial into the body. A number of organisations and individuals have expressed concerns about these developments, including uncertainties about the effects of new materials on human health and whether the types of applications envisaged would be welcomed by society at large. You will consider how these issues are assessed and what policy decisions are likely to emerge in this rapidly developing field of science.

Comparing the topics

As you progress through the course, you will be asked to make comparisons between the topics, reflecting on what you have learnt. The following section begins this process.

- ■ Having just read the descriptions of the topics, what influence do you think they might have on people's lives?

- ▨ Several topics involve science that has the potential to directly impact on the everyday lives of citizens in the UK (e.g. whether we decide to eat beef or not) and other countries (e.g. the public health crisis in Bangladesh). Equally, several of the topics investigate issues that have the potential to impact on future generations, as is the case with climate change. In this respect, scientific knowledge can inform our lives at a number of different levels, from the individual through to the global context, and in the short and long term.

These topics all involve scientists working at the frontiers of scientific knowledge, in some cases to produce novel organisms and materials, as is the case with Topic 6 (Genetic manipulation) and Topic 7 (Nanotechnology). But what scientists choose to conduct their research on is determined by a wide range of influences besides an interest in what is unknown; scientists are also very aware of what areas of science are likely to attract funding and are thought to be of importance by the scientific community as a whole. Indeed, as they conduct research, scientists draw on existing knowledge and experience from other scientists to further scientific knowledge, but this is not the only source of information. Scientists can also draw on knowledge from local communities, such as indigenous peoples, as is the case with Topic 4 (Medicinal plants).

This process of knowledge production is fundamental to science. Scientific knowledge progresses from what has been previously known or agreed through processes of investigation and verification. This process of verifying results happens at the level of individual scientists, checking and repeating findings until they are satisfied that their findings are valid, but also at the level of the wider scientific community. Publication of the details of new scientific work ensures that it is subjected to the scrutiny of other scientists, so that they can be reasonably confident that this new finding is a genuine addition to what is already known. This ongoing tension between what is and is not known is a crucial factor in the relationship between contemporary science and wider society, and will form a key background factor in much of the discussion of the topics and four themes. For example, as society has to come to terms with this ever-expanding body of knowledge, conflicts can occur in terms of novel risks and ethical issues, and how decision makers communicate these issues to the public. We will come back to this distinction between what is and is not known periodically throughout the course. What will become more evident as the course progresses is that what is known may be far from certain knowledge, in that new information can change the way old facts, theories and concepts are seen; and that a degree of scientific uncertainty will always surround what is known, to a greater or lesser extent. For example, in Topic 1 (BSE/vCJD), you will see how the interpretation of data about prion diseases was disputed by different scientists, leading to continuing uncertainty about which interpretation was (and continues to be) correct.

From this example, it is possible to see that the processes of scientific investigation and verification do not always progress effortlessly and without conflict or controversy. Scientists, like any employees, work within the demands of their profession, a key element of which is the desire to establish priority, to be the first to make a 'discovery'. What then of the processes whereby scientists achieve success? In Topic 1 (BSE/vCJD), you will consider how Stanley Prusiner overcame conflicts and controversies with other scientists about the validity of his ideas in establishing priority for his theory that prions caused BSE-like diseases. By the same token, scientists do not always get things correct at the first attempt and this can also have important implications. You will investigate issues of this nature in Topic 3 (Water and well-being: arsenic in Bangladesh).

However, contemporary science is not just about individuals working alone in laboratories. Rather, much of modern science involves scientists working collaboratively in teams, requiring state-of-the-art experimental facilities and equipment. As such, funding for contemporary science often involves large sums of money provided by a range of sources, including public bodies, charities and industry.

■ Why is funding so important to contemporary science?

▪ Funding is important because conducting contemporary science is an expensive business, requiring a range of resources, including staffing, facilities, equipment and materials.

The issue of who funds contemporary science raises ethical issues about who owns scientific knowledge (also known as intellectual property) once it is produced. You will investigate these issues further, for example in Topic 4 (Medicinal plants).

Up to now you have completed a short activity that asked you to consider what information you already knew about one of the seven topics, reflecting on where you found that information. You were then introduced to the topics in the form of short descriptions. Having read these descriptions, you have also begun the process of comparing and contrasting the different topics, particularly in terms of the ways that contemporary science is conducted. This skill of looking for similarities and differences between topics is something that you will develop as you progress through the course.

In summary, this is a course that will help you to develop an interdisciplinary perspective to contemporary science. The majority of your study time (75–80%) will be spent investigating this scientific perspective. You will also develop the skills necessary to make sense of the scientific topics in relation to four themes: communication, risk, ethical issues and decision making. This will account for 20–25% of your study time. The following sections briefly introduce each of the themes in turn. Once you have studied this material, you will be prepared to begin your study of Topic 1 (BSE/vCJD).

The course themes

This section introduces the four themes, providing definitions of key terms and noting the types of theme-based questions that you will encounter in your study of the course. You will revisit the themes periodically in relation to each of the seven topics, for instance through activities that ask you to identify a particular theme or themes and then complete a task using the information you have gathered. This can be quite challenging at first so we will also help you to identify key examples where the themes are discussed in the study text. To do this, we have adopted marginal icons for each of the themes:

C   Communication   R   Risk

E   Ethical issues   D   Decision making

It is important to note, of course, that the exploration of these themes will be far from exhaustive. Instead, it is helpful to think of the discussion of the themes as providing insights into some of the issues that help to shape our understanding of contemporary science and how it circulates in society.

Communication

C

Communication affects our everyday lives in often prosaic but also very profound ways. At a very basic level, communication involves the exchange of information between a producer (e.g. a person speaking or author of a book) and an audience (e.g. the listener or reader) via some form of message (e.g. a conversation or book). These are examples that you will hopefully relate to from your experiences of communicating in conversations and reading materials. Of course, it is important to note that the producer and receiver need to have a level of shared understanding for the communication to have a chance of being successful; otherwise, misunderstandings may occur. But that is not the end of the story: even if we understand a message, we may not agree with it. These basic premises are also apparent when we examine science communication.

In this course, we will focus on the communication of science. Science communication is fundamental to the maintenance and further development of science. For example, without science communication there would be no way of verifying scientific knowledge. In effect, without communication, routine scientific practices such as the verification of scientific knowledge would cease to function. There would also be no way of informing the public of the latest advances in scientific knowledge. As a result, the public would be both unaware of useful new knowledge and its implications, and less able to contribute to related debates and deliberations. Indeed, we often talk of a 'breakdown in communication' when there is a disjuncture between what is known between parties and what could have been communicated. Effective communication, then, is crucial to the production of scientific knowledge, its dissemination within the scientific community and wider society, and in engaging citizens in related deliberations.

What examples are useful to demonstrate how science is communicated?

There is a wide variety of examples of science communication relating to each of the seven topics in the course. We have therefore made selections of useful and interesting examples for you to explore further. For example, you will be asked to consider examples in which:

- scientists communicate with other scientists, for instance in the form of journal publication of research papers and presentations at scientific conferences;

- science communication involves other experts and professionals, for instance patent lawyers working to protect intellectual property rights, expert witnesses in public inquiries, and politicians and officials communicating science policy;

- scientists communicate with the public, for instance through media reporting (newspapers, television news and documentaries), in exhibitions and museums, in films, popular science books and magazines, and though health education campaigns;

- non-scientists communicate information relevant to the topics, for instance in the form of activists protesting against the developmental testing of GM crops, or indigenous peoples' knowledge of medicinal plants;

- information and communication technology (ICT) is used to communicate science, for instance through electronic mail (email), in the form of large data sets of digital information, as computer simulations, and as web pages on the Internet.

Taken together, these examples will illustrate how scientific knowledge circulates within the scientific community and wider society.

How is scientific information communicated among scientists, decision makers and the general public?

As you progress through the course topics, you will also be asked to consider some of the factors that influence how scientific information is communicated among scientists, decision makers and the general public. For example, you will be introduced to some of the factors that influence the selection and representation of science news for mass media outlets, including newspapers, television and radio, noting the importance of these media for disseminating newly published scientific information to a wider audience. By the same token, you will encounter examples such as public inquiries, patent applications and legal proceedings, noting the similarities and differences in how scientific evidence is represented in these examples.

One important issue that you will revisit throughout the course is that of scientists communicating new scientific information to other scientists, in particular through publication of research papers in peer-reviewed scientific journals. These acts of communicating through peer-reviewed scientific journals allow scientists to be confident that new scientific information is credible and reliable. Why? The answer lies in the process of **peer review**.

Peer review normally begins with the editors of the journal to which the manuscript has been submitted. They recruit anonymous expert reviewers who read and comment on draft copies of the submitted manuscript. These reviewers are independent of the authors who submitted the manuscript, although they are often aware of who produced the work they are checking. The real benefits of having colleagues check other scientists' work is that they can be chosen as experts in that particular field of inquiry. As such, they are much more likely to spot mistakes. If they do spot a mistake, they can request that the authors make revisions prior to acceptance for publication. Alternatively, if they feel that the paper is not of a sufficient standard, they could recommend that it be rejected. By convention, work that has successfully passed peer review is normally accepted for publication. In this way, the act of communicating through a peer-reviewed scientific journal means that the audience (mainly other scientists, but also some media professionals, including science journalists) can be confident that this new knowledge is credible and reliable.

Of course, this assumes that scientists will submit their work for peer review. For many scientists this is the normal route for communicating new research findings. Indeed, university-based research scientists in the UK, particularly those funded by the public purse, are actively encouraged to publish their work. Having said this, scientists can also discuss their work at scientific conferences prior to publication. Moreover, there are rare examples where scientists have communicated their research findings directly with media professionals (e.g. in press conferences) or through the Internet, but prior to peer review. You will consider what the implications of such approaches can be when you study Topic 6 (Genetic manipulation).

This ignores scientists who are working for, or funded by, industry or the military. Although these scientists can also submit their work for peer review, they may have conflicting contractual obligations or priorities. For example, scientists funded by industry or private money may be required to defer publication while intellectual property rights, for instance through **patents**, are secured. By contrast, scientists working for the military may be restricted in what they can publish for reasons of national security. In this way, scientists can both be motivated, for instance to gain prestige and promotion, and constrained, for instance by contractual obligations, in the ways they communicate science. This can influence whether, when, how and to whom science is communicated.

What evidence is there to illustrate that science communication is influential?

As you encounter the various topics in the course, you will investigate what types of evidence are available to demonstrate the impact of science communication on the intended audience. These types of evidence will depend on a number of factors, not least the means of communication and how many people formed the audience. In a face-to-face conversation between two scientists, for example, this could be judged on the reaction of the person hearing the communication. If one scientist asks another to take a measurement during an experiment and they do this, then the communication could be said to have been successful, but not all science communication is so easy to assess.

Even when producers do witness their audience in real time, as would be the case when scientists present their work at a conference, the effects on the audience can be very hard to judge. This becomes harder still with mass audiences, such as television audiences. In part, this is because mass audiences consist of individuals with a range of knowledge, experience, attitudes and beliefs. In these kinds of mass communication, the audience is also separated from the producer: a television newsreader is not present when the audience watches their broadcast.

What is clear is that science communication can have important effects on audiences, but these effects can be very difficult to predict. We can judge this through consideration of the findings of research that investigates audience reactions, for instance to media reporting of BSE and vCJD, and in Topic 1 you will do just that. The impact of science communication can also be assessed though indirect measures, however. For example, during the BSE/vCJD episode, consumption of British beef fell dramatically. It would be a reasonable assumption to infer that this was the result, in part, of science communication which discussed the dangers of eating BSE-infected beef. Many members of the UK public would have seen these communications and decided to avoid eating British beef, at least until they considered themselves to be at no more risk than they would have been prior to the BSE/vCJD episode. To be confident that this inference was correct, of course, would require further research.

But how could scientists and the public know what the dangers of eating BSE-infected beef were? The answer has to do with risk, which is the next theme to be introduced.

Risk

R   If we think in broad terms, risk is something that impinges on all our everyday lives. Indeed, all human activities have risks associated with them. Let's consider an example where Sukhwinder, a 21-year-old art student, decides to travel by public transport to the Tate Modern Art Gallery in London. Sukhwinder begins her journey from Manchester travelling to London on the train and then the London Underground. Both journeys have very small risks associated with train derailment. Thankfully, she arrives safely at Blackfriars Underground Station and walks across Blackfriars Bridge towards the gallery. In so doing, there is a very small risk that she might become a victim of crime. On arrival at the gallery, again unharmed, she heads for Marcel Duchamp's exhibit, *Fountain*, which consists of a men's urinal laid on its back with the artist's pseudonym 'R. Mutt 1917' inscribed on its side. Originally produced in 1917, this artwork still retains the power to shock; it could be seen as risky because it challenges the conventions of what constitutes art.

In this brief and simplified example, Sukhwinder has encountered several different types of risk: derailment, crime and artistic shock. We encounter these types of risk all the time; they are unavoidable in many cases, although we can take certain measures to reduce our exposure or mitigate the effects, for instance by hiding valuable possessions and taking out insurance, even by avoiding modern art galleries! What has this got to do with science, you might reasonably ask? Well, imagine Sukhwinder decided to buy her favourite brand of bottled water at the gallery because she was thirsty.

■ Could Sukhwinder be sure that drinking this water was completely safe?

▪ Absolute safety is unachievable. Instead of thinking of absolute safety, it is more realistic to think of relative safety or relative risk.

Sukhwinder could be fairly confident that, assuming the seal on the bottle was intact, the water would not cause her significant harm. She could base this decision on her previous knowledge and experience of drinking this particular brand of bottled water, but this is not the only information to hand. Bottled water often comes with a label which provides information about the brand, the water source and details of the mineral content, for example. Also, she has seen this brand of water advertised on the television and this has increased her confidence that the water will not make her ill. As a result, her perception of the risks associated with this product has been influenced, in this case through advertising.

Citizens living in the UK have regular access to clean, uncontaminated water. Such luxuries are not universal, however, as you will see when studying Topic 3 (Water and well-being: arsenic in Bangladesh). By studying Topic 3 you will further investigate issues of relative risk that can have profound consequences for human health.

What aspects of risk are covered in the course?

In terms of the way we have used the concept in this course, risk consists partly of the **probability** of an event coming to pass, but also involves the nature of the consequences arising from the event. Probability is something that you will address in several of the topics in the course. We have included a revision box below that outlines the key aspects of probability that you need to be aware of at this point in the course.

Revision of probability

The concept of probability is a purely theoretical one. In practice, a given experiment to quantify probability can only determine the proportion of times a particular outcome occurs in a finite number of attempts. Hence, *in theory* if we toss a coin an infinite number of times then the probability of tossing heads will be equal to that of tails, one in two, which can be represented as a fraction: ½. *In practice*, however, this distribution could be skewed in favour of either heads or tails, because we can only toss the coin a finite number of times.

Assessments of probability are a routine part of scientific investigation for many scientists. They are important because it is not possible to predict with any certainty what the outcome of an event will be. Scientists therefore use assessments of probability to demonstrate the relative likelihood of an outcome coming to pass.

If a process, such as the tossing of a coin, is repeated in identical fashion a very large number of times, then the probability of a given outcome can be defined as the fraction of the results corresponding to that particular outcome:

$$\text{probability of a given outcome} = \frac{\text{number of times that outcome occurs}}{\text{total number of outcomes}}$$

This equation shows that probabilities have values between 0 and 1 (inclusive), where a probability of 0 indicates impossibility, 0.5 represents a probability of one in two of an outcome coming to pass (such as in the tossing of a coin) and a probability of 1 indicates certainty.

Probabilities can be represented in a number of different ways, as a fraction, a decimal number, or a percentage. Hence, the theoretical probability of tossing heads could be represented in the following ways:

the probability of tossing heads is ½;

the probability of the coin landing heads up is 0.5;

there is a 50% probability that the coin will land heads up; or

there is a one-in-two chance that the coin will land heads up.

As an illustration of a risk that you will investigate in the course, consider the issue of water contaminated with arsenic that you will study in detail in Topic 3. The citizens of Bangladesh (and surrounding area) are subject to a particular **hazard**, that of arsenic in the water and its potential to harm humans. Hazards are an important aspect of risk: they are defined as the potential to cause harm. The potential to be harmed is influenced by the level of **exposure**, that is, the concentration of arsenic in the water and the amount of contaminated water that is drunk by a given individual. The greater the concentration and the longer that someone is exposed, the greater the potential to be harmed. This can be judged by investigating the **dose**, which is the amount of arsenic that reaches the organ where it presents a hazard.

Hence, if we assume that the level of arsenic contamination remains constant in a given source of water, then a Bangladeshi citizen who frequently drinks water from the source that is contaminated with arsenic is *theoretically* more likely to suffer the consequences of arsenic poisoning than a visitor to the region who drinks the water from the same source over the period of only a week. In this respect, the Bangladeshi citizen is at greater risk than the visitor because his or her exposure is higher and this is likely to raise the dose ingested. It is important to note, however, that in *practice* both of these individuals could be affected by the hazard of arsenic poisoning.

Of course, not all risks have short-term consequences. Indeed, it is possible that a risk could involve deferred harm in that someone could take a risk knowing that in all likelihood they would not be affected by that risk until later in life; for example, someone who decides to start smoking knowing that this could increase their likelihood of ill health later in life. In this way, risks can be difficult to predict because they involve dynamic factors (e.g. the level of exposure could be affected by environmental conditions and the effect of the dose could be influenced by differences between humans) that introduce uncertainties in how different individuals might be affected by the same hazard.

The example of the Bangladeshi citizen and the visitor is an example of a 'naturally occurring' risk; the water contains arsenic because of geochemical processes. It is also an example of an 'involuntary risk' in that both the visitor and the citizen have little choice but to drink the water or suffer from dehydration. If we contrast this with a 'voluntary risk' (e.g. a person choosing to eat one foodstuff over another) of a type that is 'human-induced' (e.g. foodstuffs containing novel GM materials), then this distinction is clear because the person choosing has alternatives, such as the choice of a non-GM equivalent. (However, this assumes that, in this second example, the person choosing is provided with sufficient information to make an informed choice; and to ensure that a choice is informed requires that the person understands this information.) This is an important distinction to make because research into public perceptions of risk suggests that people are more willing to accept voluntary risks than involuntary ones. Of course, assuming that an individual has all the necessary information to make an informed choice, then a naturally occurring risk could also be voluntary. Indeed, you may have been witness to a severe storm warning broadcast during a winter weather report and nevertheless still decided to travel.

Risks, then, vary according to the nature of the hazard, which is the possible source of harm, and hazards vary according to the nature of the consequences. For example, if a very large near-Earth object (NEO) collided with the Earth, the hazard would be extremely large, potentially altering climatic conditions. However, the theoretical probability of such an event coming to pass at a given point in time is extremely small. By contrast, small NEOs regularly enter the Earth's atmosphere in the form of meteors (which disintegrate as they pass through the atmosphere) and meteorites (which fall to Earth). These meteorites present a small hazard to the human race as a whole. However, if you were unlucky enough to be hit by one, as a dog was in Egypt in 1911, then the consequences could be fatal and therefore extremely hazardous. Thankfully though, there are no known instances where a human has been killed by a falling meteorite.

How are risks identified?

Scientists investigate risk by attempting to quantify the probability of a hazard coming to pass. This process will vary according to the nature of the risk involved. For example, in Topic 2 you will investigate a naturally occurring risk, that of an NEO colliding with the Earth. In identifying the risks associated with this phenomenon, scientists have used a range of data. These include evidence from previous NEO collisions (with both the Earth and the Moon) and ongoing tracking of NEOs. As with many scientific investigations, the more data that can be collected (e.g. in tracking the orbit of an NEO), the more confident scientists can be in their interpretations. This is because the increase in data reduces the level of scientific uncertainty. However, scientists cannot always accurately measure all aspects of a given risk. For instance, it is currently impossible to predict the magnitude of an earthquake in a particular place prior to its occurrence. (However, it might be possible to predict the likelihood of an earthquake taking place in a particular location at some indeterminate time in the future. Hence, if you lived close to the San Andreas Fault you would be much more likely to experience an earthquake with a large magnitude than if you lived in Glasgow.)

For human-induced risks, the process of risk identification can be slightly different. For example, scientists working with new GM or nanotechnology-based materials are potentially dealing with greater levels of scientific uncertainty because they are developing materials that have not previously been in existence. Developing measures to identify risks is therefore more difficult because there may be little or no previous evidence outside of controlled laboratory conditions to work with and any associated risks may not manifest themselves in the short term. These risks could therefore be much harder to manage in the 'real world', which brings us to the next question you will consider when studying the theme of risk.

How do you assess and manage risk?

It is important to attempt to assess risk and then seek to manage it. In the simplest of examples, risk can be represented in terms that use measures of probability (see revision box above). However, it is important to note that no matter how reliable this process of objective risk assessment might be, it is not the end of the story. Indeed, individuals may consider a given risk to be of greater or lesser magnitude for a range of reasons that can involve any combination of objective and subjective reasoning.

■ Name an example that you have encountered in the course that combines subjective and objective risk assessment.

▨ Sukhwinder combined aspects of objective risk assessment by using the information and advice contained on the bottle of water with subjective risk assessment, choosing to drink this brand of bottled water, in part, because it had appeared in television advertising and this increased her confidence that the water was safe to drink.

Measurement of risk is often based on statistical methods of prediction, whose origins lie with the insurance industry. The modern insurance industry relies on actuarial data to make predictions about risk, based on statistical calculations. Having developed statistical methods and collected actuarial data, the insurance industry now offers differentiated premiums based on the likelihood of the insurance company being required to pay out on a claim. Statistical predictions of this nature are now commonplace with experts compiling and analysing vast amounts of data for insurance companies, but they are also used by scientists. For example, scientists investigating the effects of climate change use complex computer models that are based on statistical methods of prediction.

Once identified, the data from measurements of risk inform the management of those risks and efforts to introduce risk reduction and prevention measures. And it is often the case that the more accurate the measurement of a given risk, the more likely it is that effective measures can be taken to mitigate it. This is a crucial point because scientific investigations often involve large uncertainties and those involved in managing those risks often have to make these decisions in the light of incomplete or conflicting data. Of course, this also assumes that the risk is seen to be of sufficient concern to justify mitigation, that another risk might not be given a higher priority, and that the cost of mitigating the risk is seen to be warranted.

Ethical issues

If assessments of risk examine the prospects for 'what could happen?', then
ethical issues consider 'what should happen?' Ethical issues are an important
aspect of contemporary science because they have the potential to influence
actions; for instance, how society and individual citizens respond to science.
Ethical issues are also a factor in shaping which scientific practices and
investigations are considered to be acceptable and desirable. Given recent
developments in the life sciences (e.g. genetic manipulation) and physical
sciences (e.g. nanotechnology), this is particularly pertinent. New knowledge in
these areas has the potential to raise novel ethical questions, or renew calls to
address unresolved ethical issues from similar fields of inquiry. In this way, the
ethical boundaries governing scientific practices can develop with the introduction
of new scientific knowledge as values evolve, but ethical issues may also
influence the direction of research.

Which ethical issues will be covered in this course?

You will encounter a range of ethical issues as you study S250. Here, we briefly
introduce several of those to be addressed. You have already encountered one
key ethical issue: who owns scientific knowledge? Increasingly, scientific
investigations are either exclusively or partly financed by private sources of
funding, sometimes in combination with public money from research councils,
government departments, etc. Is it therefore reasonable for society to expect that
the results of these investigations will be freely available to all?

You have also considered how scientists use existing knowledge in their
investigations, but who owns knowledge that already exists? Can new medicines,
crops and materials that draw on indigenous knowledge or materials be owned by
private companies? These questions have important implications for all science,
but particularly for scientific work that investigates medical treatments.

Increases in scientific knowledge of human genetics raises questions about who
should have access to that knowledge. Should genetic knowledge be made
available to insurance companies so that more accurate predictions of individual
risk can be made? These advances in knowledge are no less challenging when
applied to human embryos. When does human life begin, and can experiments on
human embryos ever be ethically acceptable? Is it ethically acceptable to screen
an embryo for therapeutic reasons, for instance, to create a child in part to treat
an existing child? Should parents be given the choice to alter aspects of their
unborn child's genetic make-up if it might reduce the risk of ill health in later life?
What then if the technology was sufficiently robust to select for hair colour?

A further important ethical issue relates to the themes of communication, risk and
decision making. For example, scientific investigations into NEOs are ongoing.
With these investigations comes new knowledge of NEOs and their orbits in the
Solar System, but considerable uncertainties often remain in terms of the
probability of these objects colliding with the Earth. Given the potentially
catastrophic outcomes of an NEO impact with the Earth, at what point should
scientists inform decision makers and the public of a possible collision?

Ethical issues have the potential to influence our everyday behaviour. As 'responsible citizens', many people now recycle glass bottles, newspapers and the like, and local councils have targets for levels of recycling. Increasingly, individuals are embracing the concept of 'ethical consumerism' which involves the purchase of products from sustainable sources, organic food from local sources and alternative sources of energy. As a result, funding for scientific investigations into these issues has increased; society has influenced the direction of scientific research.

How might the purposes of scientific investigation be judged as an ethical issue?

As scientific knowledge develops, society needs to take account of this new knowledge. It is often the case that ethical concerns become more prominent when the formal system of regulation, through legislation for example, is not deemed sufficient to deal with the implications of science, or when members of society feel disempowered. Scientific investigation is about the accumulation of new knowledge, but for what purpose and are those purposes ethically acceptable? If we value the pursuit of knowledge for its own sake, then this purpose could be used to justify an investigation, regardless of *why* it is carried out. (Of course, this ignores *how* the research would be conducted – see the processes for scientific investigation below.) An alternative approach would be to investigate a phenomenon for a particular purpose, such as the development of a treatment or a new material.

Both approaches have their strengths and limitations. For example, the argument of pursuing knowledge for knowledge's sake could be used to justify investigations that might otherwise be seen as ethically questionable. It could be argued that developing an astronaut-led mission to the planet Mars is justified because humans have yet to stand on the planet's surface. However, this argument could be questioned ethically in terms of the cost of such a mission when this funding could be diverted to searching for a solution to the problem of arsenic in the drinking water of millions of Bangladeshi citizens. In this sense, ethical arguments about purpose can involve judgements about how many people might benefit from the research. Moreover, pursuing knowledge for its own sake could also result in unanticipated consequences, either positive (finding new resources on Mars) or negative (returning Martian contamination to the Earth's environment). By contrast, in conducting an investigation for a particular reason, scientists could make an ethical case for conducting their investigations. The danger with this approach is that in searching for a specific purpose, other related issues may be overlooked. For example, in searching for a drug that is effective against a particular disease, such as lung cancer, a scientist might fail to appreciate the value of that drug for other conditions, such as heart disease. To address this tension, a balance needs to be struck that both limits what scientists can do in pursuing a particular purpose, but also allows some flexibility to investigate novel avenues in search of that goal. Of course, you should also bear in mind that commercial pressures shape the processes of scientific investigation. The issue of purposes should also be considered in the light of questions about who benefits, therefore, and how much can they afford to pay? These are questions you will consider in Topic 6 (Genetic manipulation).

How might the processes of scientific investigation be judged as an ethical issue?

In considering the purposes of scientific investigation, decisions also need to be made about *how* the research should be conducted. In this way, the purposes (ends) and processes (means) of scientific investigation are linked. Indeed, in many of the examples you will investigate in the course, the means and ends will be closely linked. Assuming that an ethically acceptable *purpose* has been agreed, then an ethically acceptable *process* needs to be found. For example, in the case of using animal experiments in the development of medicines it could be argued that the purpose (producing medicines) is justified but that animal testing (the process) is not. For the research to be allowed then, scientists need to address concerns about the process, for example through consideration of the three Rs of refinement, reduction and replacement originally devised by William Russell and Rex Burch in 1959.

The three Rs provide important guiding principles for those working with animal experiments, informing contemporary laboratory practices in the UK. They are defined as:

- Refinement – minimising suffering and distress to the animals involved in a given research project.
- Reduction – minimising the number of animals to be used in a given research project.
- Replacement – avoiding the use of living animals altogether by introducing alternative means for a given research project.

Of course, even if the three Rs are carefully addressed there are those who may still disagree with the use of animals in experiments on ethical grounds. The challenge for those making decisions about whether to allow a particular process to proceed has to be made in the light of a number of factors, not least whether the *processes* and *purposes* are ethically acceptable; not an easy task.

The important point to remember is that if a suitable means cannot be found, then although the purpose might be ethically acceptable, the research might not go ahead because the purposes (ends) cannot be justified by the process (means). But that may not be the end of the story. For example, the scientists conducting these investigations could choose to relocate their work to somewhere where the legal and regulatory framework is based on a different ethical value system, for example, one that allows animal experiments.

In the UK, the processes of scientific investigations are subject to various pieces of legislation, both national and international, some of which devolve regulatory powers to statutory bodies such as the Human Fertilisation and Embryology Authority. These bodies review the ethical procedures of any proposed (or ongoing) work and, where the relevant ethical issues have been addressed sufficiently, issue licences to allow the investigation to proceed. Failure to comply with the conditions of the licence could result in censure and the removal of the right to continue the investigation.

Decision making

D Scientific knowledge has the potential to affect the course of human civilisation and the environment within which we live. It is not the only factor influencing such change, but it can be a profoundly important one. It is precisely because scientific knowledge can make a difference to so many peoples' lives that decision making related to scientific issues is often the subject of considerable negotiation and contest. Decisions about whether to allow the commercial planting of novel genetically modified organisms in the UK or to reduce global carbon emissions are important precisely because they matter; they have the potential to change the way we live. Of course, not every decision made about a scientific issue will have such profound implications. Choosing to eat beef is an individual decision, but one that could be informed by scientific information about BSE and vCJD. If you consider the results of all those individual decisions together, however, as industry, local councils, regional authorities, national governments and international and global bodies often do, then you realise that individual decisions about science-related issues have the potential to directly or indirectly influence decisions made at these different levels through the development of policy. In this way, decision making about science occurs at a number of different levels, involving a range of individuals, institutions and organisations.

Who are the decision makers within each of the topics in the course?

There are a large number of decision makers involved with each of the topics in the course, many of whom are engaged in developing policy at the regional, national, international and global levels, and others who make decisions at the individual and local levels. Indeed, there are too many to examine every one. For each topic we have selected important examples, including:

- scientists and scientific institutions, for example scientific representatives from the Intergovernmental Panel on Climate Change (IPCC), scientists working on international and nationally based projects, NASA, UK research councils and the Royal Society;
- medical professionals, including those working for the World Health Organisation (WHO), general practitioners (GPs), pathologists and vets;
- media professionals, for example, journalists, editors and public relations officers;
- non-governmental organisations (NGOs), such as Greenpeace, and charities such as the Wellcome Trust;
- representatives from industry, for example, multinational pharmaceutical companies, the biotechnology industry, local farmers and 'green' businesses promoting sustainability;
- politicians and officials, for example, political representatives from the IPCC, the European Parliament, the UK Science Minister, Chief Medical Officer and Chief Scientific Adviser, and local councillors;
- other professionals and experts, such as patent lawyers, religious leaders, ethicists, and social scientists studying the relationship between science and the public;
- regulatory bodies, such as the Food Standards Agency (FSA) and the Human

Fertilization and Embryology Authority;

- indigenous peoples, such as the Xhomeni tribe who live in the Kalahari region in Africa;
- patients and their families, including those suffering from vCJD in the UK and Bangladeshi citizens suffering from arsenic poisoning;
- citizens/consumers who do not fall into any of the categories above.

Don't worry if you are unfamiliar with some of the decision makers listed above. You will be introduced to them as you study each topic.

At first glance, the categories listed above look very neat. Of course, the reality is more complex as these categories can overlap, influencing different levels of decision making. For example, the parent who is happy for their child to eat beef, the activist who campaigns against the commercial production of genetically manipulated crops in the UK and the professor of geology who gives the keynote address at a prestigious international conference could all be the same person. In this way, we are all citizens of science, whether we are a scientist, a government minister or a call centre operative, and we regularly make decisions that affect ourselves, our families and friends.

What measures can be taken to protect the public?

Given the previous discussion of the importance of identifying and measuring risk, it would seem appropriate to consider what measures can be taken to protect citizens. As we have seen, decisions related to risk are often made in a climate of scientific uncertainty. As a result, decision makers often have to make choices in the light of incomplete, partial or contested knowledge. This introduces the issue of timing; when is the optimum time to make a decision? For example, decision makers may need to consider whether to delay a decision to wait for the results of scientific research to be forthcoming. And this can take time, particularly if the research is in its early stages. Even then, the results of the investigation may be inconclusive or contested and the decision makers may find that they are in no better position to make a judgement. But that is not all. Decision makers also need to take account of existing regulatory frameworks, ethical issues, risks to the public and environment, as well as ensuring that calls for scientific progress and economic considerations (how much a particular course of action might cost in financial terms) are addressed. And then they need to consider how to keep the public engaged with these developments as well as considering the 'human costs' of a particular decision (e.g. slaughtering a herd of cows could eliminate a farmer's livelihood). As you will see from your study of the various topics, the resulting context for decision making about scientific issues is often complex.

Developed in response to concerns about human impact on the global environment, which gained momentum in the 1960s, the **precautionary principle** is an important guiding principle that informs decision making about some of the complex contemporary scientific topics that you will encounter in the course, for example, in Topic 5. The precautionary principle, as defined in the World Charter of Nature adopted by the UN General Assembly in 1982, states that:

lack of scientific certainty should not be used as a reason for postponing measures to prevent suspected or threatened environmental damage.

The precautionary principle has been included, with minor revisions, in various treaties and conventions governing environmental issues. In essence, it codifies a fundamental shift in decision making about risk, requiring that, in the absence of full scientific evidence, a precautionary approach to risk should be adopted. In practice, the precautionary principle has proved extraordinarily challenging to implement because it has been interpreted in different ways by various decision makers. However, imagine if this principle were to be *widely* adopted in decision making about contemporary science, which although it is apparent in some of the topics in the course is currently not the case for *all* scientific investigations. This would have profound implications for decision making about science-related issues because it requires that parties who are involved in research and development need to provide evidence that their actions will not introduce novel risks or cause irreversible damage.

What role does the public play in decision making about science?

Given the often contested nature of decision making described above, you may be wondering how you could participate as someone with an interest in science; as a scientific citizen. This is a challenging question to answer, particularly when we consider scientific issues discussed at the global level, such as climate change. In this instance, just ensuring that representatives from across the globe meet to discuss climate change can be a very real challenge. Of course, that does not preclude decision making at the other levels, not least in terms of how national governments approach this issue, or how you manage your own life. But what opportunities are there for members of the public to influence wider deliberations about science?

Citizens who live in democratic societies have a range of options available to them should they wish to influence decision-making processes, the most obvious being to vote in local, regional, national and, in the case of the European Parliament, international elections. Alternatively, citizens can adopt lobbying strategies or participate in legal protests, for instance as a member of an NGO involved in direct action or writing to their MP. More recently, policy makers have decided that increasing levels of dialogue and consultation between scientists and citizens could re-engage the public in deliberations about science. Several methods have been developed, mainly in continental Western Europe, to facilitate dialogue and consultation, for instance through forms of deliberative democracy including citizen juries and consensus conferences. You will examine an example of deliberative democracy in action in Topic 6 (Genetic manipulation).

However, deliberative democracy is not the only issue related to concerns about the relationship between science and society. The issue of **informed consent** also has an important role to play because it devolves the deliberative process down to those directly affected by the decisions. In effect, if sufficiently transparent, credible, up-to-date information is provided to citizens, an 'involuntary risk' can be transformed, at least in principle, to a 'voluntary risk'. For example, adult patients are provided with a range of information about their medicine and this information is an important factor informing their choice over whether to take it. More formally, patients receiving treatments can be asked to sign consent forms to acknowledge that they were aware that they may be subject to significant risks. In these ways, informed consent requires those affected to take responsibility for how scientific information impacts on their everyday lives.

Summary

1   This section has introduced a number of key issues relevant to your study of S250. Initially, you were asked to reflect on what you already knew about one of the seven topics addressed in the course, noting where you found this information. This introductory activity was devised to illustrate that scientific information has the potential to circulate widely in society and that the topics were chosen, in part, because of their prominence as issues of public discussion. In future, you will no doubt come across other scientific issues that generate public debate and we hope that the skills you develop as you study will be useful in investigating these topics.

2   You were also briefly introduced to each of the seven topics, noting the distinction between what scientists currently do and do not know about them. The importance of this distinction will become increasingly apparent as you study each of the seven scientific topics in relation to the four themes. There are a number of similarities and differences between the topics, some of which were outlined following these descriptions. You will be asked to consider further similarities and differences between the topics as you continue to study the course.

3   This section has also provided an introduction to the four themes that permeate the course. As you progress through the course, you will develop your knowledge and understanding of the seven scientific topics and how they relate to the four themes of communication, risk, ethical issues and decision making. In investigating these themes, you will be asked to consider the motivations to conduct science, but also some of the constraints and regulations whereby decisions are made about contemporary science.

4   Primarily though, this is a course about scientific facts, concepts, theories and principles and you will spend most of your study time learning about the scientific aspects of the seven topics. Topic 1 begins this process with the first of the seven scientific topics: BSE and vCJD.

Questions

Test your understanding of the *Introduction to the course* by tackling the following questions, the answers to which are found at the back of the book.

Question (i)

(a) Name the process whereby new scientific knowledge is verified prior to publication. (b) Briefly outline how this process works in practice.

Question (ii)

You will learn about near-Earth objects (NEOs) in Topic 2. (a) Are NEOs examples of naturally occurring or human-induced risks? (b) Describe the likelihood of a large NEO collision with the Earth in the next 10 years.

Question (iii)

Assume that you are a scientist preparing a research proposal that *might* involve experiments using live animals. (a) Name the three guiding principles that should inform the preparation of your bid. (b) Briefly outline how each of these could influence your research.

Question (iv)

(a) Which important principle can be used to inform decision making about complex science-based issues? (b) Briefly outline how the implementation of this principle influences decision-making processes.

Learning Outcomes for the *Introduction to the course*

S250's Learning Outcomes are listed in the *Course Guide* under three categories: Knowledge and understanding (Kn1–Kn6), Cognitive skills (C1–C5) and Key skills (Ky1–Ky6). This section outlines how these overall learning outcomes have been treated in the context of the *Introduction to the course*. (*Note:* At this early stage of the course, you are not expected to have developed new cognitive and key skills in investigating science in context; the learning outcomes for this section therefore relate to knowledge and understanding.)

Having completed the first activity, you should be more aware of the knowledge and understanding you already hold about at least one of the topics addressed in the course (Kn1 and 2).

You were also introduced to the four themes, seeing how science can be communicated among scientists, decision makers and the public (Kn3), and how subjective and objective aspects of the assessment, communication and management of risk can be important for contemporary scientific issues (Kn4). Furthermore, you have been introduced to some of the ethical issues that you will encounter during the course, noting the importance of considering the purposes and processes of scientific investigation (Kn5). Finally, you were introduced to the theme of decision making with respect to contemporary science, noting the role that scientists, practitioners working in other disciplines and members of the public can make to decision-making processes (Kn6).

1.1 Introduction

Having introduced the course's four themes – communication, risk, ethical issues and decision making – in the *Introduction to the course*, we now turn our attention to the first of seven scientific topics: **BSE** (**bovine spongiform encephalopathy**, a cattle disease) and **vCJD** (**variant Creutzfeldt–Jakob disease** in humans). These two diseases are treated together in order to emphasise the similarity of the underlying science and because the first is believed to have given rise to the second.

BSE and vCJD are important in their own right, having had major impacts on the lives of many people. Some people have died of vCJD and their deaths will have profoundly affected those who knew them. Large numbers of cattle have died either directly or indirectly because of BSE and this has had enormous economic effects on the agriculture and food industries. As a result, many practices in these industries are profoundly different from those of two decades ago. Although fundamental research into similar diseases was already well underway – leading to the development of an entirely new branch of science called **prion** biology – this research acquired a new urgency with the emergence of BSE and vCJD. It is quite likely that prion biology will lead in unexpected directions during the lifetime of this course. However, the main reason for including this topic in *Science in Context* is that, arguably, the BSE/vCJD experience has had a profound influence on the attitude of the public (at least in the UK and elsewhere in Europe) to several proposed science-based developments – in particular, the rapidly developing fields of genetic manipulation of organisms (Topic 6) and nanotechnology (Topic 7).

Activity 1.1 (Part 1)

Ongoing through Chapter 1

In order to achieve some of the learning outcomes of this course, it is important that you constantly bear in mind the four themes as you study the science underpinning a particular topic. What do we mean by 'bear in mind'? As you work through the diverse scientific topics included in the course, you should increasingly be able to analyse such topics critically from the perspectives of communication, risk, ethical issues and decision making. However, the first step is simply to be able to *recognise* these themes in the context of a particular scientific topic and the second is to be able to *explain* in what way(s) the material is relevant to one or more of the themes.

As discussed in the *Introduction to the course*, we have adopted marginal icons for each of the themes. These will be used in the early part of the course to draw your attention to material that is particularly relevant to one or more of the themes. However, in preparation for the stage where these

C R E D

icons will not be used, for the rest of this chapter we will use the icons for communication, risk and decision making – but not that for ethical issues. Instead, *you* should write the letter Є in the margin adjacent to text that you consider to be particularly relevant to ethical issues. You should also make a brief note to remind yourself in what way(s) you consider the material to be relevant. We will return to Activity 1.1 at the end of Section 1.2 in order to review progress and then again at the end of Chapter 1.

1.1.1 BSE

C

In December 1984, it was noticed that a cow on a Sussex farm was displaying head tremors and loss of coordination. The animal died in February 1985. The vet regarded this case as sufficiently serious for a post-mortem examination to be necessary. In September 1985, a government pathologist confirmed that the cause of its death was a type of disease known as a **spongiform encephalopathy**. Spongiform encephalopathies (i.e. 'spongy brain diseases') are so called because, on post-mortem examination, brain cells can be seen under the light microscope to contain fluid-filled cavities giving them a spongy appearance (see Figure 1.4). Other cattle showed similar symptoms and, in a paper published in the peer reviewed scientific journal *The Veterinary Record* in 1987, bovine spongiform encephalopathy (BSE) was formally recognised as a new disease in cattle.

Є

Because BSE is invariably fatal, cattle displaying its symptoms are usually killed without delay. Affected animals lose weight and the milk yield declines in dairy cattle. However, the most conspicuous and distinctive symptoms of the disease are walking extremely awkwardly, increased nervousness and altered behaviour or temperament (such as kicking during milking). BSE therefore soon became known in the news media and elsewhere as 'mad cow disease'.

(By the way, did you place an 'Є' in the margin alongside the above paragraph? Although cattle displaying symptoms of BSE are usually killed immediately, this is not a policy applied to humans who contract fatal diseases. This could therefore be described as an ethical issue.)

C

■ Why might journalists and others choose to refer to BSE as 'mad cow disease' rather than use the disease's proper scientific name? Suggest some advantages and disadvantages of this approach to communication.

▨ The expression 'bovine spongiform encephalopathy' doesn't exactly trip off the tongue! Furthermore, it includes neither of the keywords – 'cow' and 'disease' – that would aid effective communication with non-specialists because of shared understanding of the terms. In the early days of the disease, the abbreviation 'BSE' would have been relatively unfamiliar and so could not normally have been used without explanation. The phrase 'mad cow disease' is dramatic, memorable and clearly indicates that a cattle disease is being discussed. Within a short while, it had gained such currency that reports tended to explain that mad cow disease was also known as BSE rather than the other way around. The ensuing loss of scientific precision probably didn't matter too much – at least until we started to read of 'the human form of mad cow disease'.

During the late 1980s and early 1990s, BSE developed to epidemic proportions and considerable economic significance in the UK. The number of cases per annum grew quite rapidly for several years before steadily declining (Figure 1.1) as increasingly severe precautionary measures were taken to control the disease. (In Chapter 2, we will examine some of these control measures and their effectiveness.) By 2004, a total of over 180 000 BSE cases had been reported.

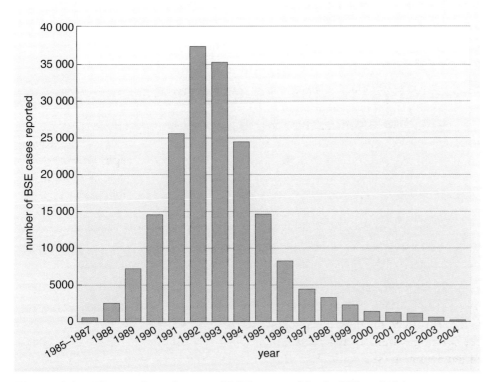

Figure 1.1   The number of cases of BSE reported in the UK to 2004.

■   In which year did the number of BSE cases reported in the UK peak?

▨   1992.

■   Approximately how many BSE cases were reported in that year?

▨   37 000.

In due course, BSE also appeared – and occasionally still continues to appear – in other countries. By mid-2004, 24 countries (Figure 1.2) had reported cases. Moreover, scientists working for the European Commission (EC) believed that BSE is 'highly likely' in eight more countries and 'cannot be excluded' from seven others. Many other countries have not allowed the EC scientists to assess the likelihood that they too have BSE. Nevertheless, BSE is unlikely to flare up again as a major problem in cattle anywhere in the world *where vigilance is maintained*.

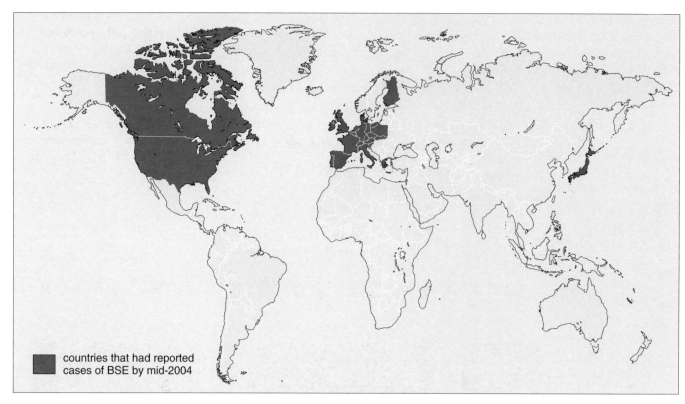

Figure 1.2   Countries that had at least one reported case of BSE by mid-2004.

Spongiform encephalopathies are often referred to as **transmissible spongiform encephalopathy (TSE)** diseases. Transmissible means that they can cause disease only if they gain access into a new host (e.g. through injection or by being ingested). In contrast, contagious diseases require physical contact between animals and, strictly speaking, infectious diseases can be transferred only by the airborne route.

1.1.2 Pre-existing human TSEs

Before BSE was recognised, several distinct human TSEs were already known, the most significant of which are outlined below. Human TSEs can have **incubation periods** of several decades between initiation of the disease and recognition of its early symptoms. Following provisional diagnosis, all these TSEs inevitably culminate in the patient's death after varying periods of decline. Definitive diagnosis required post-mortem examination of the brain. Although in all TSEs brain tissue has a pathological and usually spongy appearance similar to that seen in BSE, the different human TSEs affect different parts of the brain (Figure 1.3) and this influences the symptoms observed. Figure 1.4 shows the microscopic appearance of human cerebral cortex in individuals who have died of CJD.

[Handwritten margin notes:]

transmissible diseases are transferred by injection or ingestion.

contagious diseases are transferred by physical contact.

infectious diseases are transferred through the air.

TSE diseases can have long incubation periods. and are usually fatal

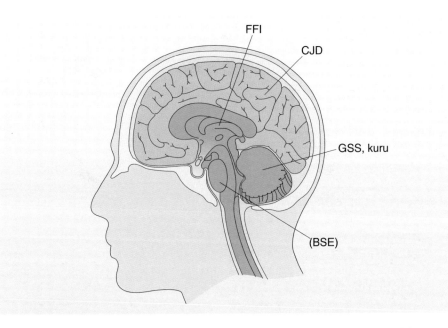

Figure 1.3   Different human TSEs mainly affect different parts of the brain: Creutzfeldt–Jakob disease (CJD), the cerebral cortex; Gerstmann–Sträussler–Scheinker syndrome (GSS) and kuru, the cerebellum; fatal familial insomnia (FFI), the thalamus. (In BSE, the main effect is on the brainstem.)

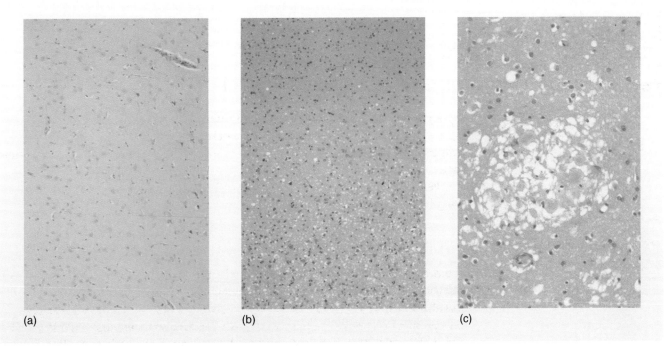

Figure 1.4   (a) Thin section through a normal human cerebral cortex, stained with a red dye. (b) A stained, thin section through the cerebral cortex of an individual with the original form of CJD; note the widespread spongiform changes. (c) A stained, thin section through the cerebral cortex of an individual with vCJD, showing local spongiform changes.

An important human TSE is **Creutzfeldt–Jakob disease (CJD)**, the annual incidence of which throughout the world is approximately 1 case per million of the population.

R

■ The current population of the UK is about 60 million. How many CJD cases would therefore be expected in a typical year in the UK?

▨ About 60. This does emphasise the comparative rarity of CJD.

R

The fact that the annual incidence of CJD among Libyan Jews is about 25 cases per million suggested that there might be something fundamentally different about the development of CJD in this group. Intriguing clusters of cases (in either time or space) have been reported from Slovakia, Hungary, England, the USA and Chile. Since in 85–90% of CJD cases no specific cause can be identified, these are referred to as **sporadic CJD**. Typically, victims of sporadic CJD are in their 50s or 60s and die within a year of onset of the illness (although they may have been incubating the disease for much longer). Symptoms include muscular spasms, dementia, loss of higher brain functions and behavioural abnormalities. **Inherited** or **familial CJD** is another form of the disease which accounts for the remaining 10–15% of cases. Worldwide, about 100 families are known to carry one of the genetic mutations responsible for it. In the past, CJD was acquired occasionally by transmission as a consequence of medical procedures involving biological material (e.g. concentrated human growth hormone or transplanted corneas) derived from people with undiagnosed CJD. This form is known as **iatrogenic CJD** (which literally means 'CJD caused by the doctor').

E

Gerstmann–Sträussler–Scheinker syndrome (GSS) is a dementia known to be genetically inherited in about 50 families worldwide. Although GSS is otherwise similar to CJD, the age of onset is more variable and so is the duration of the disease (two to six years). Worldwide, about 10 families are known to carry a genetic mutation that gives rise to **fatal familial insomnia (FFI)**, in which death occurs about one year after the onset of complete insomnia and other symptoms. Clearly, GSS and FFI are even rarer than CJD. However, detailed genetic studies of these families have contributed to our understanding of the cause(s) of human TSEs.

Kuru is a TSE disease that was formerly quite common in the Foré people of the Eastern Highlands of Papua New Guinea. Kuru's symptoms include uncoordinated movement, paralysis and an irrational laughter, which gave rise to the disease's alternative name, 'the laughing death'. Dementia is uncommon in kuru (in marked contrast to CJD). Death usually occurs within 12 months.

R

■ The annual disease-specific mortality from kuru was about 3%. How much more common was kuru among the Foré people than CJD worldwide?

▨ 3% is equivalent to 3 in 100 or 0.03. The annual incidence of CJD worldwide is 1 in a million, which is equivalent to 0.000 001. Dividing the former by the latter tells us that kuru was about 30 000 times more common among the Foré than CJD is on a worldwide basis.

Kuru affected mainly women, and children of both sexes. In fact, at one stage most deaths in women were caused by this disease and men came to outnumber women by 3 to 1. The few men who died from the disease had probably contracted it when young. Although never witnessed by outsiders, the Foré reportedly held mortuary feasts in which they ate their dead as a mark of respect. Women and children were believed to have eaten the ground-up and heated brain of the dead tribal member as a kind of grey soup, while men ate the muscle tissue. This difference presumably reflected differences in status within the social group. Kuru began to decline in the mid-1950s, after mortuary feasts were banned by the Australian authorities which then governed Papua New Guinea. By this time, the American virologist Carleton Gajdusek and others had worked out that kuru must have been contracted by eating infected human brain tissue. For this work, Gajdusek was awarded the Nobel Prize in Physiology or Medicine in 1976.

[handwritten: E]

[handwritten margin note: kuru caused by eating infected brain tissue.]

1.1.3 TSEs and non-human animals

Several TSEs of non-human animals were also known before the recognition of BSE and others have come to light subsequently. The most significant of the former is **scrapie**, a disease of sheep that has been known for over 200 years. Its symptoms include irritability, excitability, restlessness, scratching, biting, rubbing of the skin (hence its name), loss of wool, weight loss, weakness of the hindquarters and sometimes impaired vision. Some breeds are relatively resistant to the disease (e.g. Scottish Blackface) and others are much more susceptible (e.g. Herdwick, Suffolk), suggesting a genetic component. The export of sheep from Britain in the 19th century is thought to have caused scrapie to spread to many other countries. However, strict quarantine procedures seem to have prevented the disease reaching Australia and New Zealand. Although some people believe that scrapie may be becoming more prevalent in the UK, the statistics kept on the disease have generally been so poor that it is impossible to be sure. It is likely that several distinct strains of scrapie exist among sheep.

Transmissible mink encephalopathy (**TME**) is a disease first reported from a Wisconsin mink farm in 1947, but subsequently found in Canada and Finland as well. Although TME is quite rare, all the mink on a farm are usually affected in any particular outbreak. This suggests that the disease is caused by eating infected sheep or cattle carcasses, although the prevalence of fighting and cannibalism among young mink has also been implicated.

Chronic wasting disease (**CWD**) is a TSE of mule deer and elk discovered more recently in North America. **Feline spongiform encephalopathy** (**FSE**), affecting domestic cats, and the rather sweepingly named **zoological spongiform encephalopathy**, affecting a range of animals kept in zoos (e.g. antelopes such as eland, nyala, Arabian oryx, greater kudu, gemsbok; cats such as cheetah, puma, ocelot; and possibly ostrich), are both thought to have the same cause as BSE in cattle.

[handwritten margin note: Transmissible spongiform encephalopathy found in other animals besides cattle and humans.
It is found in:
sheep.
mink.
mule deer & Elk
zoo animals eg antelopes, cats & ostrich
domestic cats.

scrapie has been around for over 200 years.
spread to other countries through the export of sheep]

1.1.4 BSE and risks to human health: vCJD

C R D

Government concerned about the effect of BSE on humans - confident it could not be transferred even though it was a new disease as scrapie had been around so long without posing any threat.

Given this background, it is not surprising that the possibility that BSE in cattle *might* pose a health risk to humans was given serious consideration from a very early stage in the outbreak. Various precautionary measures intended to eradicate BSE in cattle and also to prevent any possibility of transmission of the disease to humans were introduced. We will discuss these in more detail in Chapter 2. At the same time, the public was repeatedly assured by both officials (e.g. the Chief Medical Officer or CMO) and politicians (e.g. the Secretary of State for Agriculture, Fisheries and Food) that it was perfectly 'safe' to eat British beef (see Chapter 2). One of the reasons for this confidence was that humans had not contracted any TSEs from scrapie-infected sheep in more than 200 years. Although it was generally assumed that BSE would pose no greater threat to human health than had scrapie, it should be borne in mind that BSE was a new disease, the characteristics of which were by definition unknown.

Nevertheless, following diagnosis during 1994 and 1995 of 10 cases of CJD in people under 42 years of age, variant (formerly, new variant) Creutzfeldt–Jakob disease (vCJD) was recognised as a *new* TSE in its own right. Although vCJD shares some clinical symptoms with other types of CJD, there are important differences. For instance, vCJD affects younger people (the average age at death is less than 30 years), has a longer duration (up to two years between onset and death) and has different early clinical symptoms (psychiatric or behavioural changes, such as depression, rather than dementia) than classical CJD.

C

The Secretary of State for Health stated in the House of Commons on 20 March 1996 that:

> There remains no scientific proof that BSE can be transmitted to man by beef, but the [scientific advisory] committee have concluded the most likely explanation at present is that these cases are linked to exposure to BSE before […] 1989.

By the end of 2004, there had been 148 deaths in the UK from confirmed or probable vCJD. Although current calculations suggest that about 200 UK residents will have died from this disease by the time the outbreak comes to an end, it will be many years before we know for sure. At one stage it was feared that the death toll would be *very* much higher than this (Section 1.5). There have also been a few cases of vCJD in other countries.

What then is the cause of TSEs, such as BSE and vCJD? To answer this question, we need to look at a relatively new area of scientific investigation: the biology of prions.

1.2 The biology of prions

The increasing interest in kuru during the 1950s and 1960s had the effect of stimulating research into TSEs in humans and other animals.

■ Summarise, in general terms, the possible causes of disease in animals.

A disease might have a genetic basis. Alternatively, it might be caused by a harmful agent of some kind entering the animal's body through its lungs, in its food or drink or by penetrating its skin. Such an agent might be chemical or biological.

[handwritten note: diseases may be caused by genetics or chemical or biological agents entering the body.]

A genetic mutation in the DNA of either all the cells in the animal's body (i.e. a congenital disease) or in some of them (e.g. in many cancers) may result in the production of protein that is either non-functional or does not function properly. Such a protein might be a key enzyme in a biochemical process or it might regulate the expression of other genes. Among the biological agents that cause diseases are viruses, bacteria, fungi, protoctists and small animals such as parasitic worms and insects.

[handwritten note: C]

Stanley Prusiner of the University of California at San Francisco started to research the biology of TSEs following the death of a CJD patient in 1972. Ten years later, he published a key scientific research paper in the prestigious peer-reviewed journal *Science* (see Box 1.1).

[handwritten note: C]

Box 1.1 Referencing sources of information

The author(s) of a scientific research paper (or other scholarly work) conventionally acknowledge any non-original information or ideas they use – at least, any that are not so well established as to be included in standard textbooks – by briefly citing the source in their paper and then giving full details in a list of references at the end. Thus, the author(s) might write something like 'Prusiner (1982) claimed …' or 'It has been claimed (Prusiner, 1982) …' and then in the references section of their paper give the following details of Prusiner's paper:

Prusiner, S. B. (1982) Novel proteinaceous infectious particles cause scrapie, *Science*, **216**, pp.136–144.

This gives (in order): the author's name (plus initials), the date of publication, the title of the paper, the name of the journal in which it was published (in *italics*), the volume number of the journal (in **bold**) and the paper's page numbers. This should be sufficient information to locate it.

The details of a book used as a source may be provided in the following format:

Ridley, R. M. and Baker, H. F. (1998) *Fatal Protein: The story of CJD, BSE and other prion diseases*, Oxford, Oxford University Press.

This gives (in order): the authors' names, the date of publication, the title of the book, the place of publication and the name of the publisher.

In fact, for referencing both scientific papers and books, several alternative formats are commonly used. The important thing is that the author(s) should select an appropriate format for (say) scientific papers and then use it consistently.

Although we do not reference every piece of scientific information included in this course, some references to key scientific papers, reports and books are included – along with suggestions for more general further reading – in the 'References and further reading' section at the back of each book. Guidance as to how you should reference non-original information and ideas in assignments will be given in the assignments themselves.

Some journals are described as 'prestigious' but can scientist's work be said to be any better for having appeared in them? **C**

The journal *Science* in the USA (as well as *Nature* in the UK) is often described as a 'prestigious' journal. This suggests that other journals may be less prestigious. What accounts for this difference? *Science* and *Nature* are both read by large numbers of scientists, but more importantly these scientists are drawn from diverse scientific disciplines. This is not true of, for instance, *The Veterinary Record*, which has a much more specialist readership. Although influential in a more general sense, popular science magazines such as *New Scientist* do not publish research papers and do not involve peer review. It is the publication in *Science* and *Nature* of detailed research papers that may well be read by specialists in many other scientific disciplines – and the recognition and kudos that this can bring to their authors – that makes it so desirable to have one's work appear in these journals.

C

Prion-protein molecule containing no genetic material

In his 1982 paper, Prusiner proposed – extremely controversially and based on relatively limited experimental evidence – that both scrapie and CJD-like diseases were caused by an infectious agent consisting of a protein molecule *but no genetic material*. Previously, it had been believed that any infectious agent had to contain genetic information stored in nucleic acid (either DNA or RNA). Prusiner named this protein molecule 'prion', shorthand for '**proteinaceous infectious particle**' (although, logically, he should have called it 'proin'!).

Initially many colleagues disagreed with Prusiner's hypothesis and it was openly criticised. However this did not dissuade Prusiner to continue with his research which he continued to do away from the eyes of the media.

Despite his paper having been peer-reviewed for publication in a prestigious journal – which should have established for it a very high level of credibility among both scientists and media professionals reporting science – the initial reaction to Prusiner's hypothesis among some fellow biologists has been described as ranging 'from scepticism to outrage' (although others welcomed its explanatory power). After the American scientific magazine *Discover* roundly criticised what it perceived as his promotion of the prion hypothesis in a 1986 article tellingly entitled 'The Name of the Game is Fame: But is it Science?', Prusiner resolved not to talk to journalists while he and his colleagues concentrated on their research into the biology of prions. In 1997 he was awarded the Nobel Prize in Physiology or Medicine for his discovery of 'Prions – a new biological principle of infection'. A few experts (including Gajdusek, the recipient of the first Nobel Prize awarded for work on TSEs) remained unconvinced by the 'protein-only hypothesis' of the cause of TSEs. Nevertheless, TSEs are now often called prion diseases and increasing numbers of scientists refer to themselves as prion researchers.

eventually awarded Nobel Prize.

Gajdusek and a few others still remain sceptically.

Activity 1.2

Allow 15 minutes

The previous three paragraphs contain several hints that science may not always be conducted in the disinterested, dispassionate way in which it has traditionally been portrayed. Re-read the paragraphs carefully, looking for evidence of controversy and how this might have affected the debate. Write two or three brief paragraphs summarising this evidence and then compare these with our 'Comments' on this activity at the end of this book.

Although it is now *generally* accepted that all TSEs are caused by prion proteins as proposed by Prusiner – and that this hypothesis explains the essential features of these diseases – this was certainly not the situation when BSE arose in the mid-1980s or earlier when researchers were trying to understand how diseases such as CJD, kuru and scrapie were transmitted.

A major problem facing TSE researchers was that there seemed to be a genetic basis to some TSEs (e.g. familial CJD, GSS and FFI), whilst a biological agent of some kind seemed to be responsible for others (e.g. kuru). To complicate matters further, in yet other TSEs there seemed to be *both* a genetic *and* an infectious component (e.g. although scrapie was widely believed to be spread through sheep grazing contaminated pasture, some breeds seem to be quite susceptible to the disease whilst others seem to be relatively immune). In addition, the relatively long incubation periods of some TSEs made it difficult to identify their initial cause(s) or to study them. This resulted in considerable confusion and contest between different research teams and certainly no consensus on the underlying biology of TSEs.

At an early stage in his work on TSEs, Prusiner deliberately infected mice with scrapie to use them as **animal 'models'** of the disease (see Box 1.2). He then showed that extracts from the brains of these mice:

- caused scrapie in other mice when injected into their brains
- contained high concentrations of a particular protein with a particular three-dimensional shape, or **conformation**.

Furthermore, these brain extracts lost their infectiveness when they were exposed to treatments that destroyed proteins, e.g. protein-digesting enzymes (proteases) or short wavelength ultraviolet light, but not when they were exposed to treatments that destroyed nucleic acids, e.g. nucleic acid-digesting enzymes (nucleases) or longer wavelength ultraviolet light. At least some of this protein was the nucleic acid-free biological agent that Prusiner had called a 'proteinaceous infectious particle' or 'prion'. He went on to isolate a particular conformation of a protein that appeared to be unique to scrapie-infected brains. Because this protein was *relatively* resistant to protease enzymes, which readily degrade most proteins, he called it a **protease-resistant protein** or **PrP**. Prusiner surmised that PrP protein and prion were one and the same thing.

ε

extracts from infected mice brains caused scrapie in other mice and contained a protein with a particular conformation

Box 1.2 The use of animal 'models' to study diseases

There are several reasons why there had been rather limited progress over the years in studying scrapie in sheep. Sheep are quite large animals that normally have to be kept in fields, where they are exposed to all sorts of uncontrolled aspects of the physical and biological environment which might have a bearing on whether or not they develop scrapie. They also have relatively long generation times and normally produce only one or two offspring at a time. Thus, working with sheep is both slow and comparatively expensive. It is also difficult to achieve adequate replication and sufficient control over potentially relevant variables in experiments with these animals.

Smaller animals (such as mice, rats or hamsters) breed faster and more prolifically than sheep. Large numbers can be kept conveniently and relatively cheaply in controlled conditions in laboratories. Furthermore, after many generations of inbreeding these laboratory animals are genetically uniform, thus eliminating a potential source of variability. Thus, many of the problems of working with sheep could be by-passed by artificially infecting small laboratory animals with scrapie so that they served as experimental 'models' for scrapie-infected sheep.

Of course, Prusiner was not especially interested in scrapie for its own sake. For him, scrapie was effectively a model of the human TSE (CJD) that he was studying. However, the sort of experiments he was doing could not possibly be carried out on either human patients or volunteers even if they were fully informed of any risks. Informed consent is a legal requirement in such circumstances in the UK and most other countries.

E

Some people would object as a matter of principle to artificially infecting *any* animal with a fatal disease, even if the purpose was to understand and eventually cure that disease or a similar one in humans. In this instance, it would be the means and not the purpose to which they objected (see the *Introduction to the course*). Others might question the appropriateness of mice, rats or hamsters as 'models' for either sheep or humans. However, the facts are that through experiments like these, Prusiner and others made enormous strides in developing our understanding of TSEs. These days, laboratory animals can be genetically engineered to produce particular proteins of other species (such as sheep, cattle or humans) in their brains instead of their own versions of these proteins. These genetically modified animals are assumed to be even more appropriate as 'models' of other species.

We now go on to discuss TSEs in terms of molecular biology. This course assumes you are already familiar with basic molecular biology from previous studies. In case you are not confident about the terminology, Box 1.3 provides a brief outline. In order to adequately understand the biology of prions, you may have to study or revise basic molecular biology more thoroughly.

Box 1.3 Revision of basic molecular biology

In eukaryotic organisms (whose cells have nuclei, in contrast to prokaryotes such as bacteria which don't), most of the genetic information is stored in chromosomes in the nucleus. Each chromosome consists of a DNA (deoxyribonucleic acid) molecule and various proteins. DNA molecules exist as two extremely long intertwined strands of subunits called nucleotides, each consisting of a molecule of the sugar deoxyribose, a phosphate group and a nitrogenous base. Since there are just four types of base (adenine, cytosine, guanine and thymine, or A, C, G and T for short), there are four types of nucleotide. Many (but not all) genes code for proteins, such as enzymes. Proteins consist of relatively long strands of about 20 different amino acid subunits. Each individual amino acid is coded for by three consecutive nucleotides (a triplet) in one of the strands (the coding strand) of the DNA molecule. The first stage in the production of a protein molecule (transcription) involves part of the coding strand of a DNA molecule (a gene) being copied as single-stranded mRNA (messenger ribonucleic acid) molecules (Figure 1.5). The mRNA molecules leave the nucleus and enter the cell's cytoplasm. There, organelles called ribosomes attach themselves to the mRNA molecules and effectively 'read' them. Ribosomes – together with transfer RNA molecules (tRNAs) and enzymes – add amino acids to the growing protein chains according to the sequence of triplets encountered in the mRNA. This process (translation) is completed when the ribosome 'reads' a particular RNA triplet that is always interpreted as 'stop'.

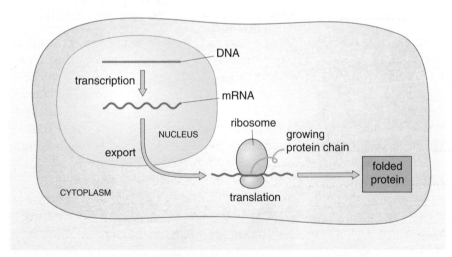

Figure 1.5 Transcription of DNA into mRNA and translation of mRNA into simple protein molecules.

Scientists established the amino acid sequence in the PrP protein, worked out the DNA sequence that gives rise to PrP and searched for that DNA sequence among the genes of mice and, in due course, people. In fact, the **PrP** gene (see Box 1.4 on the names of genes) has been found in every species of mammal so

eukaryotes have nuclei; prokaryotes don't

PrP - protease resistant protein

far investigated. When the *PrP* gene is switched on in the nucleus of a cell, PrP protein is synthesised at ribosomes in the cell's cytoplasm. Although the *PrP* gene is present in every nucleated cell of the body, it is switched on mainly in brain cells. Brain cells therefore produce lots of PrP protein. This suggests that PrP protein must play an important – but, so far, poorly understood – role in the brain. On the other hand, mice in which the *PrP* gene is 'knocked out' (i.e. rendered non-functional by genetic engineering) before birth seem to be normal apart from having problems with their daily (circadian) rhythms of sleeping, eating, etc.

c

each gene can exist in one or more forms – each form is a different allele.

Box 1.4 The names of genes

Nice though it would be simply to use one of the 26 letters of the alphabet (*A, B, C,...Z*) to refer succinctly to each gene, this is not possible. Humans alone have between 20 000 and 25 000 genes distributed around our 46 chromosomes. Considering all living organisms, there are huge numbers of different genes. The PrP protein is coded for by the gene known as *PrP* (note that, conventionally, the names of genes are italicised). The names of various other genes (e.g. *Hb* and *CPEB*) are used later in this chapter. In addition to distinguishing between different genes, it is often necessary to distinguish between different alleles (or 'versions') of the same gene. Thus, two alleles of the human haemoglobin gene are referred to as Hb^A and Hb^S.

As soon as proteins are synthesised, they fold spontaneously into complex 3-D shapes or conformations (Figure 1.5). The precise conformation into which a protein folds depends largely on the sequence of its amino acids. Moreover, a protein's behaviour within a cell is strongly influenced by its conformation. This is most clearly seen in enzymes, in which the molecules' conformation determines which reactants can be brought into contact with one another and therefore which product(s) can be produced – in other words, which reactions can be catalysed. This is the so-called 'lock-and-key' hypothesis of enzyme action which you may have met in previous studies.

conformation determines catalytic action in enzymes.

Prusiner realised that *without any change in their amino acid sequence* PrP proteins exist in (at least) two conformations. He called the 'normal' conformation **PrP^C** (for **cellular PrP**) and the abnormal conformation **PrP^Sc** (for **scrapie-causing PrP**). Compared to PrP^Sc, more of the PrP^C molecule folds into helices (the **α-helix** structure) and less folds into pleated sheets (the **β-sheet** structure) (Figure 1.6). Crucially, whilst PrP^C is soluble in cells, PrP^Sc molecules collect together into insoluble deposits. Cells containing such deposits no longer function normally and eventually die. The loss of these cells leads to holes in brain tissue, the 'spongy' effect typical of all these diseases (see Figure 1.4).

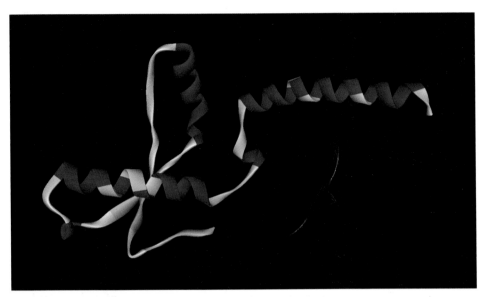

Figure 1.6 Structure of part of a 'normal' human PrP, i.e. PrP^C, protein molecule. Note the preponderance of α-helices (shown in red). The two short sections of β-sheet are shown in pale blue. (White denotes unstructured regions.) The yellow areas within the α-helical regions are amino acid positions at which mutations associated with TSEs have been found.

Figure 1.7a shows how PrP protein acts within a cell in normal circumstances. Within the nucleus, the *PrP* gene is transcribed into mRNA. The mRNA migrates out of the nucleus into the cytoplasm. At ribosomes, the mRNA is translated into a sequence of amino acids corresponding to the sequence of nucleotide triplets in the *PrP* gene. Even as it is synthesised, the growing amino acid chain folds spontaneously into the characteristic conformation of PrP^C protein (i.e. largely comprising α-helices). The PrP^C protein is then transported to the cell membrane, where it becomes attached to the cell's external surface.

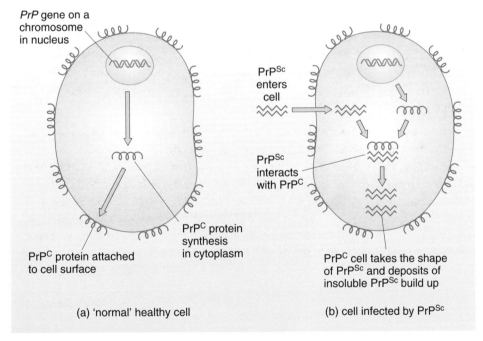

Figure 1.7 (a) How PrP protein acts within a cell in normal circumstances. (b) Prusiner's explanation for how a cell with 'normal' *PrP* genes becomes infected with PrP^{Sc} protein.

In the nucleus PrP gene is transcribed into mRNA. The mRNA leaves the nucleus and enters the cytoplasm.

Figure 1.7b summarises Prusiner's explanation for how a cell becomes infected with PrP^{Sc} protein. PrP^{C} protein is synthesised in the cell as described above. However, in this case one or more PrP^{Sc} molecules have entered the cell from elsewhere and interact in some way with newly synthesised PrP^{C} molecules in the cytoplasm. These interactions cause the PrP^{C} molecules to become PrP^{Sc} molecules by changing their conformation (i.e. by increasing the proportion of β-sheet structure compared to α-helical regions in the molecules). Not only can these newly created PrP^{Sc} molecules then clump together and disrupt the cell's normal functioning, they themselves can also interact with PrP^{C} molecules, causing the production of yet more PrP^{Sc} molecules. (Note that, unlike PrP^{C}, PrP^{Sc} does not appear on outside of the cell membrane.) It can readily be appreciated that this is a 'chain reaction' in which more and more PrP^{Sc} molecules accumulate in infected cells. Furthermore, any of these PrP^{Sc} molecules released from an infected cell (e.g. upon its death) become available to infect other cells. As more and more cells – particularly brain cells – become infected with PrP^{Sc}, the animal develops symptoms of the TSE and eventually dies. Mice in which the *PrP* gene has been 'knocked out' experimentally – and which therefore do not synthesise PrP protein – do not develop TSE diseases.

Protease resistant protein scrapie. causing proteins are transferred by injection or ingestion

Where might the PrP^{Sc} molecules that infect 'normal' individuals have come from? Clearly, in Prusiner's mice experiments they were injected into the recipient animal in the brain extracts from animals that already had scrapie. In the case of kuru, abnormal PrP molecules are presumed to have been present in the tissue – particularly the brain tissue – of people whose bodies were eaten. It is likely that kuru started from a sporadic case of CJD and became established as a relatively common disease within the Foré tribe through some of those who participated in these mortuary feasts becoming infected with kuru in this way and then themselves being eaten after death and so on. The fact that the disease has now almost disappeared some five decades after the cessation of cannibalism supports this explanation.

Activity 1.1 (Part 2)

10 minutes now, but ongoing

Go back over Sections 1.1 and 1.2, locate the places where you placed an E in the margin and compare your choices with those suggested in the 'Comments' on this activity at the end of this book, where short explanations are provided. How do these explanations compare with your own notes?

From now until the end of this chapter, continue to place an E in the margin when you identify material that you consider to be particularly relevant to ethical issues. However, this time write more-detailed explanations of the way(s) in which the material is relevant to this theme. You will be asked to compare these explanations with our 'Comments' when we return to Activity 1.1 for a final time at the end of Section 1.6.

1.3 The origin and spread of BSE

■ In the light of the above discussion about prions, what is the most probable explanation for the *spread* of BSE among cattle?

▨ The cattle presumably consumed material containing PrP^{Sc} protein.

Nevertheless, other routes of transmission could not be ruled out and therefore had to be investigated systematically. The most obvious of these are cow-to-calf transmission (either through genetic inheritance or direct contact) or cow-to-cow transmission (either through direct contact or via some aspect of their common environment such as the fields they share).

Soon after BSE was first recognised, an initial study was commenced into the pattern of BSE's spread within and between *populations* of cattle – that is, the disease's **epidemiology**. At the same time, of course, other scientists were studying the detailed biology of the disease in *individual* animals. Epidemiology can throw light on the cause(s) of a disease, how it is spread and ultimately on the effectiveness of various measures introduced to control it. On the basis of this initial epidemiological study, veterinary scientists had concluded by December 1987 that the BSE epidemic had been set off by the inclusion in protein-rich **concentrated cattle feed** (generally referred to simply as '**concentrates**') of **meat and bone meal (MBM)** derived from scrapie-infected sheep. Concentrates are fed to dairy calves, which are taken from their mothers soon after birth so that the cows can be milked. Adult dairy cattle are also given concentrates at times when their energy demand exceeds that available from grass (e.g. during winter). Once their milk yield started to decline (at about five-and-a-half years of age), dairy cows were slaughtered and their meat used in cheaper meat products such as pies, burgers and sausages. In contrast, the calves of beef cattle are allowed to suckle from their mothers for several months (which is why they are often referred to as 'beef suckler cattle') and are seldom given concentrates. These calves are then reared for one or two years before being slaughtered for meat.

Although, concentrates have always consisted mainly of plant material, for some time before BSE arose protein from almost any source was included provided it was sufficiently cheap. The feet, brains, intestines, lungs and excess fat from all animals killed in abattoirs was treated to separate the fat from the residual solid material – a process known as rendering – and the solid residue ground up and sold as MBM. It must be emphasised that, although few people unconnected with farming and the food industry would have had detailed knowledge of these procedures, they were perfectly legal at that time and were not regarded as unsafe.

It is not universally accepted that the *initial* source of infection was material from scrapie-infected sheep in concentrates. An alternative view is that BSE arose spontaneously in one or a small number of cattle, tissues from which ended up in MBM. Nevertheless, once BSE began to spread, increasing amounts of the MBM fed to cattle would have been derived from BSE-infected cattle. This would have enabled BSE to spread even further and faster. Thus, from an early stage the view of scientists advising the government was that BSE was caused by contamination of MBM derived from **ruminant animals** (ruminants include

E.

dairy cattle and their calves are feed 'concentrates'

beef cattle calves are very rarely given concentrates

dairy cattle are slaughtered at about 5½ years and can be used in cheaper meat products

D beef cattle are slaughtered between 1 and 2 years

either from scrapie-infected sheep or from cattle spontaneously infected with BSE.

47

Alterations in the rendering process.

cattle, sheep and goats). However, animal tissues had been included in cattle feed for several decades prior to the mid-1980s and scrapie had been present in UK sheep for more than 200 years (whilst BSE could have arisen spontaneously in cattle at any time). Did anything occur at about this time that might explain why BSE started then? In fact, in 1980 the government allowed a relaxation in the regulations controlling rendering. Previously, it had been a batch process in which waste animal material was heated to remove water and fat, then any residual fat dissolved in a hydrocarbon solvent and finally steam at 100–120 °C was used to remove the solvent. After the change in regulations, many rendering plants went over to a continuous process in which dry heat was applied to remove water and fat from the material, with the hydrocarbon solvent extraction stage being omitted.

■ Summarise the main changes to the rendering process that occurred following the change in regulations.

▨ The use of hydrocarbon solvents in a wet, low-fat environment was replaced by the application of dry heat in a high-fat environment. Furthermore, the original batch process was likely to have taken longer than the 'more efficient' continuous process that replaced it.

Probably because the market price of fat had fallen, only two of the 46 rendering plants in the UK were still using solvent extraction by 1988 (Figure 1.8).

Figure 1.8 Percentage of meat and bone meal produced from 1964 to 1988 by rendering plants using solvent extraction.

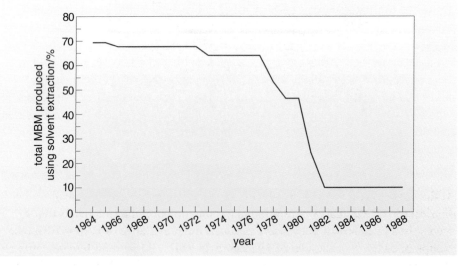

■ Cattle feed is produced in many rendering plants around the country, which supply farms in their immediate locality. Does the fact that the only two plants still using the solvent extraction process in 1988 were in Scotland relate to the geographic incidence of BSE from 1985 to 1988 shown in Figure 1.9?

▨ Yes. There were no cases of BSE in central and northern Scotland during 1985 to 1988. In southern Scotland up to 1.9% of herds were affected. In some parts of England, more than 4% of herds had cases of BSE.

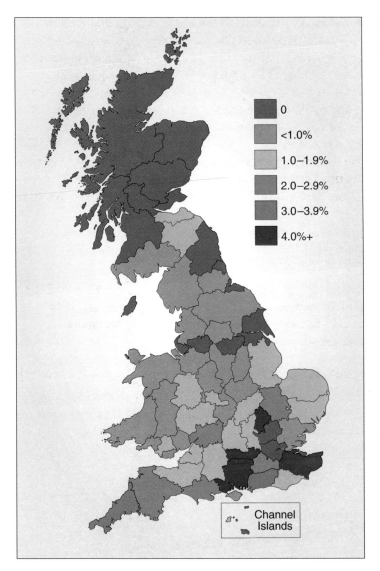

Figure 1.9 Incidence of BSE-affected herds from 1985 to 1988 as a percentage of the total number of herds by county/region.

Legend:

- 0
- <1.0%
- 1.0–1.9%
- 2.0–2.9%
- 3.0–3.9%
- 4.0%+

Channel Islands

Question 1.1

At the same time as veterinary scientists investigated the geographical distribution of BSE cases in relation to the location of rendering plants that still used solvent extraction, they also examined the prevalence of BSE in different groups of cattle. In a sample of 192 cases of BSE, 190 were in female cattle and two were in male cattle. The national herd contained 3 200 000 female and 37 000 male cattle in 1987.

(a) Expressing your answers to appropriate numbers of significant figures, calculate:

(i) the percentage of the 192 cases of BSE that involved female cattle;

(ii) the percentage of the national herd that was female.

(iii) Is there any evidence from your answers to (i) and (ii) that either sex was more prone to BSE than the other?

(b) Using the data in Table 1.1, calculate to appropriate numbers of significant figures:

(i) the proportion of BSE cases in the national herd of beef suckler cows;

(ii) the proportion of BSE cases in the national herd of dairy cows.

(iii) Express the proportion of BSE cases in dairy cows to the proportion of BSE cases in beef suckler cows as a ratio in the form $x : 1$.

(c) As mentioned above, dairy calves are usually removed from their mothers soon after birth whereas beef calves suckle from their mothers for several months. On the basis of these differences in the two cattle production systems, suggest an explanation for the answer to (b)(iii).

Table 1.1 The number of cases of BSE in beef suckler and dairy cows from 1985 to 1988 in relation to the number in the national herd.

Cow type	Number of BSE cases	Number in national herd
beef suckler	14	880 000
dairy	696	2 320 000

Thus, support for the hypothesis that contaminated cattle feed was responsible for the origin and spread of BSE was provided by:

- the geographical distribution of BSE cases in the early days of the epidemic (1985–88) in relation to the number and distribution of rendering plants still using solvent extract; and

- the relative incidence of BSE among dairy and beef suckler cattle.

R

By 1988, it was generally accepted by veterinary scientists that one or more of the post-1980 changes – but most probably the elimination of solvent extraction – had caused the rendering process to be less effective at deactivating any scrapie or BSE agent present in the MBM.

■ Of course, prions were not widely discussed at this time. But how would the above conclusions be expressed today in terms of prion biology?

■ The new continuous process was in some way less effective than the old batch process at deactivating PrP^{Sc} protein present in animal material sent for rendering prior to its inclusion in MBM. This material included tissue from either scrapie-infected sheep or cattle in which BSE had arisen spontaneously. However, once the BSE outbreak got underway, this material certainly included tissue from BSE-infected cattle. Cattle then consumed concentrates that incorporated MBM containing PrP^{Sc} protein. Within the cells – particularly the brain cells – of these cattle, interaction between this PrP^{Sc} protein and the 'normal' PrP^{C} protein of these animals caused some of the latter to be converted into PrP^{Sc} protein. The increasing amounts of PrP^{Sc} protein in the brain cells of these cattle caused many of them to develop the TSE that became known as BSE. Furthermore, ever-increasing amounts of PrP^{Sc} protein became included in MBM – and hence in concentrates – which caused BSE to be transmitted to even more animals.

New rendering process less effective at deactivating PrPSc protein found in animal material included in Meat and bone meal.

An important issue that has not been fully addressed so far is the cause(s) of infective TSEs, such as BSE and kuru. This is clearly relevant to the origin of vCJD in humans. However, we first need to consider the causes of some of the TSEs that are known to be inherited. Box 1.5 revises a few more aspects of basic molecular biology.

Box 1.5 Revision of basic molecular biology

As we have seen, the particular amino acid occupying each position in a protein is coded for by three consecutive nucleotides (a triplet) in the coding strand of the DNA molecule. Some amino acids are uniquely specified by one DNA triplet (e.g. methionine by TAC). Others are specified by several alternative triplets (e.g. valine by CAA, CAC, CAG and CAT). These relationships form part of the genetic code (which is usually expressed in terms of nucleotide triplets in mRNA rather than in DNA).

In the case of many genes in eukaryotes, enzymes remove triplets from newly synthesised mRNA molecules before they leave the nucleus; this is part of the process called post-transcriptional modification. The newly synthesised protein molecules also undergo processing. Post-translational modifications of newly synthesised protein molecules in the cytoplasm include the removal of certain amino acids and the attachment of sugar side-chains to others in order for the protein to become functional.

Although the human *PrP* gene (located on chromosome 20) comprises 253 triplets, those at positions 1–22 and 231–253 are not represented by amino acids in the 'mature' PrP protein because of post-translational processing. Some variation is possible in the triplets 23–230 without rendering the PrP protein completely inactive. However, certain mutations at particular triplets are associated with various TSEs. For instance, a mutation in triplet 102 that causes the amino acid proline to be replaced by leucine is linked to GSS. Similarly, a mutation in triplet 200 that causes glutamine to be replaced by lysine is linked to the form of CJD that is particularly prevalent among Libyan Jews. The CJD clusters reported from Slovakia, Hungary, England, the USA and Chile are also now all believed to be due to mutations at triplet 200. A combination of the triplet that codes for methionine rather than the one that codes for valine at position 129 and that which codes for asparagine rather than aspartic acid at position 178 is linked to FFI. (Some of the mutation sites in the human PrP protein are shown in Figure 1.6.)

The phrase 'is linked to' was used in the previous paragraph because these less common genotypes might enhance the rate of spontaneous conversion of PrP^C protein to PrP^{Sc} protein or they might increase an animal's susceptibility to infection by PrP^{Sc} protein from elsewhere (e.g. in food). It would be too simplistic to say that a particular mutation 'causes' a particular TSE, because whether or not the disease develops almost always depends to some extent on the environment – both internal and external.

■ Bearing in mind the above information about the genetics of some inherited TSEs, what might trigger sporadic CJD?

[handwritten margin notes:]

Base pairing rules in transcription

DNA base	RNA base
A	U
G	C
C	G
T	A

mutations at different triplets give rise to the production of different amino acids which are associated with various TSE's. (can't say they actually cause them).

- An individual's *PrP* gene might code for PrP^C protein that (1) spontaneously converts to PrP^{Sc} protein particularly easily or (2) is particularly susceptible to conversion to PrP^{Sc} through interaction with PrP^{Sc} from an external source.

So, the human *PrP* gene certainly displays some genetic variation (see Box 1.6 for a brief revision of basic genetics terminology). The various genotypes give rise to several phenotypes with respect to the PrP protein. These phenotypes appear to differ mainly in the ease with which the PrP protein changes from being comparatively rich in α-helices (PrP^C) to being comparatively rich in β-sheets (PrP^{Sc}), either spontaneously or as a result of coming into contact with PrP^{Sc} protein from elsewhere. Given this variation *within* a single species, it is reasonable to expect there to be some systematic differences in the *PrP* gene – and hence the PrP protein – *between* different species of mammal. If there are such differences in the amino acids sequences of typical PrPs in sheep, cattle and humans, a number of important questions arise. Can sheep PrP^{Sc} effect the conversion of cattle PrP^C into cattle PrP^{Sc} and, if so, how easily? Similarly, can sheep and/or cattle PrP^{Sc} effect the conversion of human PrP^C into human PrP^{Sc} and, if so, how easily? These questions relate to the existence of possible **species barriers** between the current host species of a prion disease and potential new host species.

Box 1.6 Revision of basic genetics terminology

In the context of genetics, the appearance (and also internal anatomy, biochemistry, behaviour, etc.) of an organism is referred to as its phenotype. Thus, blue and brown are alternative phenotypes for human eye colour. Many phenotypes are determined partly by an organism's environment and partly by its genetic make-up or genotype. Diploid sexually reproducing species have two sets of chromosomes – and therefore two sets of genes (except for those on the sex chromosomes, i.e. X and Y in humans) – one set derived from the mother and one set from the father. Many genes exist as several alternative alleles. For instance, there are three common alleles (*A*, *B* and *O*) of the main blood group gene. An individual possesses two copies of each gene – one maternal and one paternal. Where these copies are identical, the individual is described as being a homozygote or homozygous (e.g. the genotypes *AA*, *BB* and *OO* that give rise to the blood group phenotypes A, B and O respectively). Where the copies are not identical, the individual is described as being a heterozygote or heterozygous (e.g. the genotypes *AO*, *BO* and *AB* that give rise to the blood groups phenotypes A, B and AB respectively).

If a protein can have two alternative amino acids at a particular position along its length (i.e. different phenotypes are possible with respect to this protein), then the two genes coding for that protein present in an individual can either be different alleles or the same allele. In other words, the individual can be heterozygous or it can be homozygous for either one allele or the other.

genotype refers to the alleles
phenotype refers to appearance

■ Cite an example – discussed earlier in this chapter – in which a prion disease definitely crosses between different species of mammal.

▨ Prusiner worked with scrapie-infected rodents. Since scrapie is a disease of sheep, this is an example of a prion disease crossing between different species – albeit as the result of a human-induced experiment.

Differences in the amino acid sequence of the PrP protein of one species and that of another might make it impossible for any PrP^{Sc} of the first species to interact with PrP^C of the second species so as to convert the latter into PrP^{Sc}. In such a case, the species barrier must be regarded as insurmountable. Alternatively, the PrP proteins of two species might be sufficiently similar for PrP^{Sc} of the first species to convert PrP^C of the second species into PrP^{Sc}, but not as easily as PrP^C can be converted into PrP^{Sc} within the first species.

■ Suggest three ways in which relative 'ease of conversion' of PrP^C to PrP^{Sc} might vary.

▨ Relative 'ease of conversion' might be reflected in variation in (1) the length of the incubation period of a TSE (i.e. the time from infection with PrP^{Sc} to the first appearance of symptoms of the TSE), (2) the amount of PrP^{Sc} required to trigger development of the TSE or (3) the ways in which an animal can become infected with PrP^{Sc} (e.g. absorption of PrP^{Sc} from food might be sufficient, infection might require injection of PrP^{Sc} into the blood-stream or the only possible route of infection might be injection of PrP^{Sc} directly into the brain).

■ How would you expect the incubation periods to compare between an animal infected with PrP^{Sc} derived from a member of its own species and an animal infected with PrP^{Sc} derived from a member of another species?

▨ The incubation period might well be shorter when a species is infected with PrP^{Sc} derived from a member of its own species than when it is infected with PrP^{Sc} derived from another species.

■ Explain this possible difference in incubation period in terms of prion biology.

▨ Initially, PrP^{Sc} from one species with a particular amino acid sequence has to interact with PrP^C of another species which is likely to have a slightly different amino acid sequence. However, once some PrP^{Sc} with the second species' amino acid sequence has been produced, *this* PrP^{Sc} will probably more readily be able to convert more of the second species' identical PrP^C (i.e. with the same amino acid sequence) into PrP^{Sc}.

■ How might the amount of PrP^{Sc} required to trigger a TSE differ before and after a species barrier has been breached?

▨ A higher dose of PrP^{Sc} might be required to breach a species barrier than to transmit a TSE within a species.

Similarly, once the species barrier has been breached it might be possible to transmit a TSE within a species simply through the presence of that species' PrP^{Sc} in food. However, to breach a species barrier in the first place it might be necessary to inject 'foreign' PrP^{Sc} into a second species' blood or even directly into its brain.

It has been suggested that there are no impenetrable species barriers. Any barriers that appear to be impenetrable do so simply because the incubation period is longer than the normal lifespan of the potential new host species.

■ How could this be demonstrated experimentally?

■ PrP^{Sc} of the first species might be injected into the brain of a member of a second species. After some time, some brain tissue from the latter animal might be injected into the brain of another member of the second species. This procedure might then be repeated several more times before a member of the second species finally displays symptoms of the TSE. (*Note*: this experiment has been done.)

C D

As we have seen, the veterinary scientists who carried out the initial epidemiological study of BSE concluded that BSE is effectively scrapie that has crossed the species barrier between sheep and cattle because changes in the rendering process allowed still-infective PrP^{Sc} protein from sheep to be consumed by cattle. In fact, in its final report published in October 2000, the official BSE Inquiry (chaired by Lord Justice Phillips) came down in favour of the disease originating in cattle in South-West England during the 1970s or early 1980s. A further review (chaired by Professor Gabriel Horn), specifically into the origin of BSE, considered this suggestion to be plausible but necessarily speculative and concluded that scrapie could not be ruled out as the source of BSE. We will consider these official inquiries in more detail in Chapters 2 and 3.

There is now some concern that BSE might somehow cross back over the species barrier between cattle and sheep. Indeed, it is possible that scrapie may already be masking the presence of BSE in sheep. A major problem is that we don't know what BSE in sheep would look like – BSE, scrapie or something else? In early 2005, it was reported for the first time that BSE had been detected in a goat in France (where these animals – which are quite closely related to sheep – are commonly kept to provide milk for cheese-making) and another in Scotland.

C D

The possibility that BSE might have been triggered by the mandatory use of an organophosphate pesticide to eliminate warble fly in cattle, as suggested by organic farmer Mark Purdey and others, was also considered by the BSE Inquiry. Although the Inquiry ultimately rejected this idea, there are intriguing suggestions in Purdey's data that an imbalance between copper and manganese in nerve cells – whether reflecting the local natural environment or caused by pesticide treatment or industrial pollution – might make humans and other animals more likely to develop TSE diseases. A particularly interesting aspect of this 'story' is the great difficulty Purdey initially experienced in getting his hypothesis taken sufficiently seriously by the scientific 'establishment' to obtain research funding, etc. Effectively, he became a self-trained scientist and eventually persuaded some university-based researchers to work with him, resulting in the publication of several peer-reviewed papers.

BSE orginated from:
scrapie
infected cattle
or
triggered by
pesticides used to
control warble fly

■ Does Purdey's experience bring to mind the treatment of another TSE researcher?

▨ There are some parallels with the treatment of Stanley Prusiner, whose ideas were initially rejected by many of his peers. However, Prusiner was already an established scientist with a well-equipped university laboratory, which enabled him to pursue his research interests anyway. Ultimately, of course, he was able to convince (most of) his critics and became an 'establishment' figure himself when awarded the Nobel Prize.

Prusiner's prion hypothesis may eventually be rejected or refined beyond recognition as more is learned about TSEs or the nervous system generally. However, it is currently the most convincing explanation for the cause of TSEs and the one most widely accepted by the scientific community because it has provided a very effective foundation for further research. On the other hand, Purdey's (rather less all-embracing) hypothesis appears to be languishing on the sidelines. This contrast epitomises a difficult dilemma for science. On the one hand, there is the danger of accepting as valid an ill-founded hypothesis with all that this might entail in terms of wasted time and resources. On the other hand, there is the danger of rejecting a truly prescient hypothesis because minds are not sufficiently receptive to it. This dilemma clearly impinges on both communication and decision making.

C D

hypotheses can be rejected without consideration.

1.4 The emergence of vCJD

We now turn our attention to vCJD.

■ If vCJD really is 'the human form of BSE' (as it is often described), how is it likely to have crossed the species barrier from cattle to humans?

▨ The majority of victims probably consumed food that contained cattle PrP^{Sc} protein. The cattle PrP^{Sc} protein then interacted with their own PrP^{C} protein converting it to human PrP^{Sc}. Human PrP^{Sc} would have been much more effective than cattle PrP^{Sc} in converting more human PrP^{C} into PrP^{Sc}, which would build up in the victims' brain cells until vCJD was eventually diagnosed.

caused by ingestion of contaminated food.

■ Which parts of cattle incubating BSE are most likely to have been the source of PrP^{Sc}?

▨ The most infective part of cattle would probably have been brain tissue. Since other nerve cells are also likely to have been relatively rich sources of PrP^{Sc}, the spinal cord and the eye are also likely to have been hazardous.

D

On the basis of work on scrapie-affected sheep, various other organs were also considered to be possible sources of infection in BSE-affected cattle. In the early days of BSE, high priority was therefore given to trying to ensure that not only the head and spinal cord, but also the thymus, tonsils, spleen and intestine (and hence the lymphatic system) from any cattle that might have been incubating BSE were not incorporated into food intended for human consumption (e.g. accidentally during the recovery of as much meat as possible from the carcase)

head, spinal cord, thymus, tonsils, spleen, intestine including lymphatic system.

or recycled to other cattle in MBM. At the time, these banned parts were referred to as **specified bovine offals (SBO)**. However, since technically the head is not regarded as offal, they are now referred to as **specified bovine materials (SBM)**. In fact, only nervous tissue has been found to be infective in the case of BSE. Indeed, it has been argued that it would have been more effective to have enforced the ban on nervous tissue from cattle entering the human or cattle food chain much more rigorously than to have spread the 'safety net' as widely as was done. This is an example of how difficult it is to apply the precautionary principle in practice.

■ How else might vCJD victims have contracted the disease?

▨ As in the case of iatrogenic CJD, vCJD could have been contracted through medical procedures involving the transfer of vCJD-infected tissues.

Since prions are notoriously resistant to degradation (e.g. they remained infective after the new MBM rendering process), there was concern about the effectiveness of sterilisation of re-usable surgical instruments in hospitals – particularly after they had been used for eye or tonsil operations. On the other hand, many surgeons also had reservations about the precision they could achieve using disposable instruments. Although CJD has occasionally been transmitted via surgical instruments, by mid-2004 no known cases of vCJD could be attributed to this cause. Of course, great care also has to be taken when post-mortem examinations are carried out, to ensure that pathologists and technicians are not infected.

C R D

There was also concern about whether vCJD could be transmitted through transfusion of blood from someone incubating the disease at the time of donation. As a precaution, several countries banned blood donated in the UK for transfusion. This policy appeared to have been a sensible precaution when a UK resident died of vCJD in 2003 having in 1996 received transfusion of blood that had been donated by someone who died of vCJD in 1999. It was announced almost immediately that blood donations would no longer be accepted from anyone who had themselves received a blood donation since 1980. In 2004, a patient who died from another cause was found to have been incubating vCJD at the time of their death. This patient too had received transfusion of blood donated by someone who later died of vCJD.

could be transmitted through blood transfusions however could be that this had nothing to do with it.

■ Does it follow that the recipients of the blood transfusions necessarily contracted the disease from contaminated blood?

▨ No. It is possible that in both pairs of linked cases both the donor and the recipient of the blood contracted the disease from eating infected meat products.

However, the authorities decided even on the basis of the first pair of cases that the risk of contamination through blood transfusion was too great to ignore. The occurrence of the second pair of linked cases appears to justify this cautious approach. However, in exercising the precautionary principle, the authorities certainly went further than required by the available scientific evidence.

The death in 2004 of the second blood recipient was also significant for another reason that relates to genetics. The 142 people who had died from vCJD in the

UK up until then had all been homozygous for the triplet in the *PrP* gene encoding the amino acid methionine at position 129 (i.e. their genotype was *MM*). People with this genotype make up 40% of the UK population, with about 50% being heterozygous for methionine and valine (*MV*) and 10% homozygous for valine (*VV*) at this position. It had therefore been hoped that 60% of the UK population was immune to vCJD – or at least considerably less susceptible than those who had contracted the disease thus far.

■ Suggest another explanation for the pattern of vCJD in the UK until 2004.

▨ It might simply have been that the incubation period for vCJD is longer in people with the *MV* and/or *VV* genotypes than in those with the *MM* genotype.

The second blood recipient who was discovered to have being incubating vCJD at the time of their death in 2004 from another cause had the genotype *MV*. This suggests that the majority – and possibly all – of the UK population is potentially vulnerable to vCJD, mainly following past exposure to cattle PrP^Sc in food.

When a UK-based research team analysed the genotypes of various ethnic groups from around the world, they found that more than 75% of Foré women over the age of 50 were heterozygous for amino acid 129 of the PrP protein (i.e. their genotype was *MV*). This is a much higher percentage of heterozygotes than would be expected by chance.

■ In terms of natural selection, what does this suggest?

▨ It suggests that, during the recent past, women who were heterozygous at this triplet (*MV*) were at a selective advantage in that many of their contemporaries who were homozygous (either *MM* or *VV*) contracted kuru and died relatively young.

Heterozygote advantage is an example of balancing selection, a mechanism that maintains two (or more) alleles in a population when one would expect that, over evolutionary time, an allele that had even the slightest advantage over the other(s) would come to predominate. Probably the best known example of heterozygote advantage in humans relates to the haemoglobin alleles *Hb*^S and *Hb*^A.

■ If you have met this example of heterozygote advantage in your previous studies, try to recall how it operates. Otherwise, read the following answer.

▨ *Hb*^S*Hb*^S homozygotes suffer severe sickle-cell anaemia and usually die before reproducing. One would therefore normally expect the *Hb*^S allele to be extremely rare in a population and that most individuals would be *Hb*^A*Hb*^A homozygotes. However, in parts of Africa where malaria is endemic, the *Hb*^S allele is far more common than would be expected if it arose only through occasional mutation of the *Hb*^A allele. The explanation is that, despite suffering mild anaemia, *Hb*^A*Hb*^S heterozygotes are resistant to malaria whereas *Hb*^A*Hb*^A homozygotes are not. Thus, balancing selection – in which many *Hb*^A*Hb*^A individuals are debilitated by malaria and *Hb*^S*Hb*^S individuals die from sickle-cell anaemia – maintains *both* alleles in the population.

[handwritten margin note:] 40% population are MM. 50% population are MV. 10% population are VV.

In fact, the research team found that *MV* heterozygotes were relatively common in *all* the human populations they examined. This suggests that humans generally may have been subject to balancing selection that has maintained both alleles in the population. Controversially, the team concluded that this meant that cannibalism – one of the great taboos in modern human society – was probably widespread among our ancestors, an interpretation that was challenged by other scientists. This is an interesting example of scientific data being accepted as accurate by all parties, but with disagreement about how the data should be interpreted.

Just as it is not known whether BSE is a disease that arose spontaneously in cattle or scrapie that has crossed the species barrier, it is also not known whether kuru is a TSE that arose spontaneously in humans or one that arose in another species (most probably, pigs) and then crossed the species barrier into humans. However, because the species barrier *into* pigs is known to be very high, on balance it is likely that kuru originated as spontaneous CJD in humans. Either way, almost certainly kuru became more common among the Foré through their practice of cannibalism during mortuary feasts.

Kuru originated spontaneously.
E.

1.5 The epidemiology of vCJD

R D

In 1990, six years before the probable link between BSE and vCJD was established, the CJD Surveillance Unit was set up in Edinburgh. All suspected TSEs in humans have to be referred to this Unit, which maintains the UK's official data on all forms of these diseases. Figure 1.10 is a plot of the number of cases referred to the Unit from 1990 to 2004.

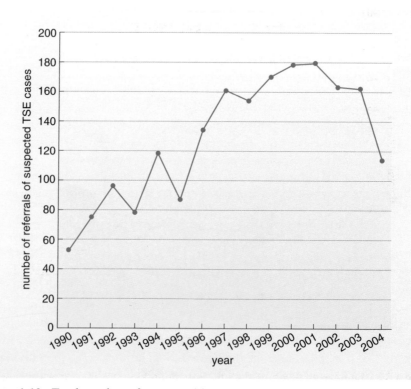

Figure 1.10 Total number of suspected human TSE cases in the UK referred to the CJD Surveillance Unit from 1990 to 2004 (as at 1 July 2005).

■ Describe any trends that you can see in these data.

▨ Although there was considerable fluctuation from year to year, the annual number of cases referred to the Unit did increase from 1990 (when there were about 50) to 2001 (when there were about 180). Since then there has been a noticeable decrease in the number of cases. This is especially evident for 2004.

■ What might account for the growth in the annual number of suspected cases referred to the CJD Surveillance Unit during the 1990s?

▨ There might have been a genuine increase in the incidence of human TSE diseases during this period – possibly attributable largely to the appearance on the scene of vCJD in the mid-1990s. Alternatively, as a result of increased awareness of BSE and vCJD, members of the medical professions may have become more alert to the possibility that some patients, whose deaths might previously have been attributed to other causes (such as Alzheimer's disease), might be suffering from a TSE disease.

The reduction in the number of suspected TSE cases reported to the CJD Surveillance Unit in 2004 might reflect either a recent decline in awareness about TSEs among members of the medical profession or greater proficiency in eliminating the possibility that a patient might have a TSE disease.

Figure 1.11 is a plot of the number deaths in the UK from definite or probable sporadic, iatrogenic, familial and variant CJD, and also GSS, from 1990 to 2004.

did disease increase or was it down to better medical awareness?

likewise is the decrease in the disease from 2004 due to a decline in medical awareness or due to precautions taken to eliminate it?

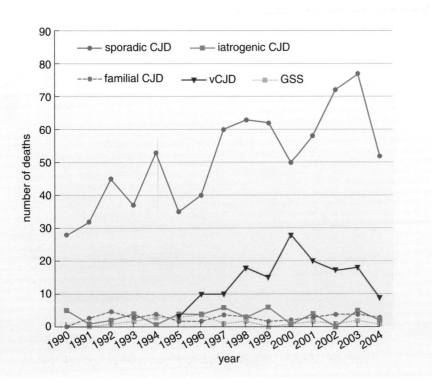

Figure 1.11   Number of deaths in the UK from definite or probable sporadic, iatrogenic, familial and variant CJD, and GSS, from 1990 to 2004 (as at 1 July 2005).

■ Do any of these data throw light on whether there has been a genuine increase in the number of human TSE cases since 1990 or whether increased reporting accounts for the apparent rise?

■ Clearly, the increase in the number of vCJD cases must have been genuine, as there were no reported deaths from this entirely new disease before 1995 and then some every year since then. There is no reason to suppose that the number of people dying of sporadic CJD has increased in recent years. Therefore, the overall upward trend in the number of deaths from sporadic CJD (from fewer than 30 in 1990 to more than 70 in each of 2002 and 2003) suggests greater reporting of suspected cases.

It is impossible to detect trends in the data for iatrogenic and familial CJD, and GSS, because of the small number of cases. The fluctuations are probably random. Nevertheless, it is rather worrying that deaths from iatrogenic CJD appear to be continuing despite all the precautions that have been taken.

■ What might deaths from iatrogenic CJD as recently as 2003 and 2004 reflect?

■ These relatively recent deaths may well reflect the typically long incubation periods of TSEs in humans. Most of these patients probably contracted the disease through medical treatments (such as the use of growth hormones derived from people with undiagnosed CJD) many years previously, before the dangers were fully appreciated.

Figure 1.12 shows the number of deaths in the UK from definite or probable vCJD from 1990 to 2004 plotted separately. The relatively sharp decline between 2003 and 2004 may be partly due to delays in confirming that patients who died in 2004 definitely did die from vCJD (which requires post-mortem examination of brain tissue) but it may indicate that the worst has passed.

Figure 1.12 Number of deaths in the UK from definite or probable vCJD from 1990 to 2004 (as at 1 July 2005).

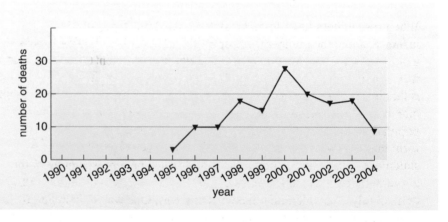

Once vCJD had been recognised and its probable link with BSE widely accepted, it became a very high priority to assess the likely magnitude of the vCJD epidemic. Epidemiologists at Imperial College London have periodically estimated the probable number of deaths in the UK from vCJD. Their 1997 **prediction** was that the disease *might* cause up to 10 million deaths. This figure was revised downwards to 50 000 in 2002 and 7000 in 2003 (with the strong likelihood that it would drop further).

■ What do you suppose was the main reason for such dramatic revisions of these predictions?

▨ Looking at Figure 1.12, it is clear that any prediction made in 1997 must have been based on very little data. Indeed, it might be argued that it was irresponsible to place such a premature prediction in the public domain. By 2002 and 2003, rather more data were available and therefore the pattern of vCJD cases with time would have been clearer.

CRD

the more knowledge gained the better predictions can be made.
with greater accuracy

■ In what senses might the later predictions be regarded as 'improvements' on the earlier ones?

▨ Hopefully, more recent predictions are more *accurate* than earlier ones. A more accurate prediction is one that is closer to the final number of cases (although, of course, only time will tell). The availability of more data would also allow the researchers to increase the *precision* of their predictions (i.e. reduce the *random uncertainty* surrounding them).

It is important to distinguish between **accuracy** and **precision**. A prediction might be accurate but known imprecisely (i.e. the true value falls somewhere within a fairly wide range of possible values). A prediction might also be precise but inaccurate (i.e. although a fairly narrow range of possible values is quoted, the true value in fact lies outside this range). Of course, normally the aim is to make predictions that are both accurate and precise (that is, with the true value falling within a fairly narrow range of possible values). In this case, the researchers' predictions were given in the form of their **best estimate** of the number of eventual deaths from vCJD together with **upper and lower 95% confidence limits** for this estimate (see Box 1.7).

Box 1.7 Statistical probability

When researchers have to make statements about a **population** – for instance, about the number (or proportion) of organisms having a particular genotype or the mean value of a particular measurement (such as height) – they generally have to employ probabilistic terminology. The reason for this is that it is usually impossible and/or undesirable to count or measure all the members of the relevant population. Instead, a **sample** is drawn from the population; the sample is then counted or measured and the data obtained form the basis for a statement about the entire population. For such a statement to have any validity, the sample must be representative of the population from which it is drawn. That is, the sample must not be either deliberately or accidentally biased in any way. One way to help guard against accidental bias is to make sure the sample size is sufficiently large that random fluctuations have little effect on the data. It can be surprising just how small a 'sufficiently large' sample can be – for instance, about 30 measurements can often adequately represent a very much larger population. A good test of the representativeness of a sample is to check that other similar-sized samples drawn from the same population produce similar data.

The same issues arise when it is necessary to make predictions about future trends on the basis of the limited information that might currently be available (as in the case of trying to predict the number of eventual vCJD deaths from the small number of cases that had occurred up to any particular point in time).

Probabilistic statements often take the form of giving for a population a best estimate (for example, of the proportion of a particular genotype, of the mean height of organisms or of the eventual total number of vCJD cases) together with an indication of how much higher or lower than this the true figure might actually be. The latter information is often provided as upper and lower 95% confidence limits. Effectively, these claim that the probability that the true value is (or will be) either greater than the upper confidence limit or less than the lower confidence limit is 5% (or, equivalently, 0.05 or 1 in 20). Of course, there is a relatively small possibility that the true value *is* outside these limits. This is the nature of probabilistic statements. The reason why 95% is often used for confidence limits is that, *conventionally*, a result expected on fewer than 5% of occasions is regarded as statistically significant. The closer the upper and lower confidence limits are to the best estimate – or the smaller the difference between the upper and lower confidence limits – the greater the precision (that is, the smaller the random uncertainty) of the statement or prediction.

In 2003 (based on data up to the end of 2001), the Imperial College team's best estimate of the number of deaths from vCJD in the UK by 2080 was 200 with upper and lower 95% confidence limits of 7000 and 10 respectively.

■ Compare this lower 95% confidence limit of 10 deaths with the number of deaths that had already occurred by the end of 2001.

▨ According to Figure 1.12, by the end of 2001 there had already been considerably more than 10 deaths from vCJD.

C   What the researchers were saying was that, although they thought that the total number of deaths would be close to 200, the number could well be either higher or lower than this. However, they believed that the probability that this number would either be greater than 7000 or less than 10 was less than 5% (0.05 or 1 in 20). The number of deaths by 2001 had already exceeded the lower 95% confidence limit. Similarly, the final number of deaths is expected to fall well short of the upper 95% confidence limit.

C   ■ In what way(s) might a reader potentially interpret a newspaper headline such as 'Up to 7000 [or 50 000 or 10 million] UK deaths from vCJD'?

▨ Expressing the prediction in this way might have the effect of encouraging readers to assume that it is *likely* that vCJD will cause the deaths of many more than 200 people.

■ What challenges can you see in reporting the researchers' predictions more
   comprehensively than implied by the above hypothetical headline?

C

▨ Eye-catching headlines are necessary to induce most people to read a
   newspaper article. While headlines should not be misleading, they must
   usually convey a simple message.

It would be difficult enough to convey the subtleties of 95% confidence limits in
the body of an article. The challenge of doing so in a headline would be even
greater. On the other hand, perhaps readers – having followed the BSE/vCJD
episode as it developed over the years – would prefer more informative headlines
such as 'Scientists predict 10–7000 vCJD deaths in UK with 95% confidence'.

The predictions discussed above are based on:

• the number of actual cases of vCJD reported – which is why their precision
   has increased over the years (i.e. their random uncertainty has decreased)

• a **mathematical model** of how vCJD spreads in the population (see Box 1.8).

Box 1.8 Mathematical and computer models

Increasingly, we read that scientists have employed mathematical models of
the real world – usually run on computers – in their research. The history of
mathematical modelling goes back a long way. Sir Isaac Newton
(1642–1727) encapsulated his law of gravitational attraction between two
bodies in a single elegant equation. These days the phenomena that
scientists are trying to understand (e.g. the Earth's climate) involve so many
interactions that they can seldom be represented nearly so simply.
Fortunately, powerful computers are now available to run models based on
multiple interacting equations – equations that would have taken many
person-years to solve not so long ago. Of course, any model incorporates
assumptions about the real world that some researchers would accept and
others might not. It is therefore incumbent upon researchers to make their
assumptions clear and to justify them. One interesting way of validating a
computer model is not just to use it to make predictions, but to use it to
make so-called retrodictions; that is, to run the model from starting
conditions that were believed to apply at some stage in the past and to see if
it predicts reasonably accurately the conditions that are known to pertain
today. We shall meet several examples of the use of mathematical and
computer models in this course.

As noted in Box 1.8, any mathematical model incorporates a number of
assumptions. A very important assumption built into the Imperial College model
was that people contracted vCJD only from eating contaminated meat.

■ What other possible routes of infection does this assumption ignore?

▨ It ignores the possibility that vCJD might be contracted from infected surgical
   instruments or through blood transfusion.

In fact, although at the time of writing (2005) there have been no cases of vCJD attributable to infected surgical instruments, we have seen that there have been two cases of vCJD that may have been contracted through transfusion of blood from donors who subsequently died of vCJD.

■ Given what was known about genetic susceptibility to vCJD when the 2003 prediction was made, what other assumption built into the model will probably have to be revised?

▨ Until 2004, it was assumed that only the 40% of the population homozygous for methionine at position 129 of the PrP protein (i.e. those that had the genotype *MM*) was susceptible to vCJD. With confirmation that a person whose genotype was *MV* had contracted vCJD (although vCJD wasn't the cause of their death), the model will have to be revised to allow for the possibility that people whose genotype is *MV* (and probably also *VV*) might develop vCJD, given sufficient time.

C R D   At present, there isn't a routine test that could be given to members of the general public to ascertain whether or not they are harbouring PrP^{Sc} protein – and therefore whether they are at risk of developing vCJD. Even if there were such a test, its use to improve estimates of the number of vCJD cases that there might eventually be would be hugely controversial. If it were possible to warn

E   people that they had a much higher than average possibility of developing the disease – and assuming they wanted to know – would it be right not to tell them? On the other hand, imagine the impact that such information would have on anyone's life.

C   In 2004, the results from tests on 13 000 preserved tonsil samples kept from tonsillectomy operations were published. These suggested that the number of vCJD cases might be around 4000 – much higher than the Imperial College team's estimate of around 200. However, this figure was an extrapolation from just three samples that tested positive for PrP^{Sc} (two of which were doubtful, a situation not uncommon in clinical diagnoses).

■ Even if all three of these tests were reliable, why should one be cautious about the figure of 4000 cases of vCJD?

▨ Extrapolation from even three reliable positive results would be problematic because of the effects of random fluctuation. Imagine the impact on the prediction of even one fewer or one additional positive result.

D   ■ There are plans to test about 100 000 fresh tonsil samples over the next few years. Suggest a reason why this survey may not throw a great deal of light on the eventual extent of the vCJD epidemic.

▨ Most tonsillectomy operations are performed on fairly young children, most of whom were likely to have been born long after there was any significant risk of eating beef contaminated with PrP^{Sc} protein.

D   ■ Suggest some reasons why reasonably reliable estimates of how the vCJD epidemic is likely to develop are needed.

■ The health service would certainly need to plan ahead if there might eventually be 50 000 victims of vCJD – let alone 10 million – rather than about 200 victims. Hard decisions might also have to be taken about how much resource to devote to trying to develop preventative treatment or a cure for vCJD depending on the likely number of victims. If the epidemic does   E. eventually fade away having claimed about 200 lives – with 148 having already died by 2004 – then no pharmaceutical company is going to embark upon the necessary research and development programme.

Although it is generally accepted that most vCJD victims contracted the disease in the mid- to late 1980s through eating contaminated meat as teenagers or young adults, there are other possibilities. One of these is that they may have been infected as early as 1970 through eating contaminated baby food. Since their gut walls are more permeable, babies may be more susceptible than adults to infection from food.

■ What additional assumptions about BSE and vCJD have to be made if this hypothesis is to be taken seriously?

■ One would have to accept that BSE existed about 15 years prior to its 'official' recognition. Further assumptions are that beef was included in baby foods at that time (which should not be too difficult to establish) and that some of this beef might have been contaminated with BSE (virtually impossible to establish after all these years). Another implication is that the incubation period of vCJD is about 25 years rather than about 10 years.

Part of the evidence cited in support of the baby food hypothesis is the decline in the annual number of vCJD cases in recent years (Figure 1.12) and the continuing relatively low average age of victims. It is argued that this might represent a decline in a 'first wave' of vCJD, rather than the disease's disappearance.

■ If so, what would comprise a possible 'second wave' of vCJD?

■ People who ate contaminated meat in the 1980s and are currently incubating vCJD, but who do not yet display any symptoms of the disease.

1.6 Is prion-like behaviour exceptional or the norm?

At the time of writing (2005), it is widely – but certainly not universally – accepted that TSEs are triggered by prions. Prions consist entirely and exclusively of PrP protein. In particular, they contain no nucleic acid – and hence no genetic information – at all. An animal may either produce its own disease-triggering PrP^{Sc} protein (in the case of inherited and probably some sporadic TSEs) or PrP^{Sc} protein from elsewhere might start a 'chain reaction' in which PrP^{C} protein synthesised by the animal may be converted into the PrP^{Sc} conformation. Three main objections to this protein-only explanation of TSEs have been put forward.

[Handwritten margin notes:]
no infectious agent

prion - proteinaceous infectious particle - misshapen protein.

proteins contain no genes - they are what genes code for - the product of genes.

ε.

The first objection was that prion biology somehow contravenes Francis Crick's 'central dogma' of biology: DNA makes RNA makes protein (which, in turn, produces the organism's phenotype).

■ Do prions contravene the 'central dogma'?

▪ No. Every PrP molecule – whether PrP^C or PrP^{Sc} – is coded for by a *PrP* gene. The amino acid sequence of the PrP molecule is specified by the sequence of nucleotides in the gene and is not altered by any change of conformation that the molecule may later undergo. (However, implicit in the central dogma is that the same protein produces the same phenotype; in the case of PrP, this is not the case.)

While this first objection can readily be dismissed as the sort of misunderstanding that tends to arise as people come to terms with new concepts in biology or any other science, the other two objections must be taken more seriously.

The second objection is that it is extremely difficult to demonstrate that TSEs are caused by purified prion protein with absolutely no involvement of another molecule (which might contain genetic information) *so small that it is as yet undetectable.*

The third objection relates to the existence of different strains of particular TSEs. For instance, there are at least 20 distinct prion strains in mice and several strains of scrapie in sheep. The *existence* of these different strains might be explained by there being different alleles of the *PrP* gene within a species. However, the *perpetuation* of these distinct strains through successive cycles of injecting infective prions into members of the same species (e.g. mice) and, particularly, into members of a different species (e.g. mice TSE transmitted to hamsters) cannot be explained in this way.

■ Suppose a TSE strain in mice was due to a particular allele of the *PrP* gene that codes for a particular sequence of amino acids in the resulting PrP protein. Explain why the perpetuation of this TSE strain through a series of mice possessing different alleles of the *PrP* gene and (especially) through a series of animals of a different species (such as hamster), following initial injection of PrP^{Sc} from the first mouse, would be surprising.

▪ The PrP^{Sc} injected initially would have the amino acid sequence of the first mouse. Any *new* PrP^{Sc} produced in the second mouse as a result of interaction with this injected PrP^{Sc} would have the amino acid sequence of the second mouse and not that of the first. In other words, the original strain of TSE would be lost as the TSE passed through the second mouse or any subsequent ones. Similarly, one would expect a hamster infected with a particular strain of mouse TSE to develop hamster TSE and not a particular strain of mouse TSE. One certainly wouldn't expect to be able to inject PrP^{Sc} from the first hamster into a second hamster and for this second hamster to develop the original strain of mouse TSE.

It is clear therefore that the existence of different prion and TSE strains cannot be explained by invoking different sequences of amino acids in PrP proteins.

■ If different TSE strains cannot be explained by differences in the genotypes of the host animals, what is the only remaining logical explanation other than prions containing at least small amounts of nucleic acid? (*Hint*: Think in terms of molecular conformation.)

■ If the existence of several different TSE strains cannot be explained by genetics, then it can only be that PrP proteins with the same sequence of amino acids can fold into several different conformations. In other words, there is more than one way in which the conformation of PrP^{Sc} differs from that of PrP^C. Furthermore, each PrP^{Sc} conformation perpetuates itself by causing any PrP^C molecule with which it interacts to adopt its own particular conformation, even if the PrP^C molecule has a slightly different amino acid sequence than its own.

Confirmation that absolutely purified prion protein is infective and that prion strains are a consequence of distinct, self-propagating conformations of otherwise identical amino acid chains came in 2004 from experimental work on the fungus *Saccharomyces cerevisiae*. Perhaps surprisingly, prions are found in several species of fungi. However, these prions are not disease-causing and may even be beneficial to the host. The protein with prion properties used in this work, called Sup35p, is concerned with terminating the synthesis at ribosomes of amino acid chains.

That the ability of several proteins to adopt different conformations may benefit their fungal hosts – and certainly not harm them – suggests that we should not necessarily regard prions exclusively as disease-causing 'rogue' molecules. Indeed, as more research is done on prion biology, it is becoming clear that the ability to acquire and transmit change in conformation may even be part of 'normal' biology. Two diverse research fields in which prion-like behaviour in proteins is implicated are cellular time-keeping and memory.

James Morré of Purdue University in West Lafayette, Indiana, has shown that proteins called ECTO-NOX (found in both animals and plants) can oscillate in unison between two conformations and that this might be the basis of time-keeping in cells. It may therefore be no coincidence that mice engineered to lack prion protein have problems maintaining daily (or circadian) rhythms.

Nobel Laureate Eric Kandel of Columbia University in New York has suggested that prion-like proteins might be responsible for 'marking' neurons prior to the laying down of more connections (or synapses) between them. Such 'marking' is believed to be the basis of memory. This suggestion was based on Kandel's own work on the sea slug *Aplysia* (whose *CPEB* gene is very similar to the mammalian *PrP* gene).

There is now little doubt that TSEs are caused by the build-up of PrP^{Sc} proteins in brain cells and that this build-up usually results from a 'chain reaction' in which contact between PrP^C protein and PrP^{Sc} protein causes the former to be converted into the latter. Scientific research into prions is also beginning to reveal that other proteins engage in similar conformation-changing behaviour and that this may be an aspect of 'normal' biology that has remained unsuspected until quite recently. Interesting developments in this field are likely to occur during the lifetime of this course.

Activity 1.1 (Part 3)

Allow 15 minutes

Compare your own notes on the relevance of material in Sections 1.3–1.6 to the theme of ethical issues with the explanations given in 'Comments on activities'. Note that, for the remainder of this topic, marginal icons will be used for all four themes.

Having covered the molecular biology and epidemiology of BSE and vCJD in this chapter, we can now examine how the BSE/vCJD episode was managed (Chapter 2).

Summary of Chapter 1

1   BSE is a TSE disease of cattle that was formally recognised in 1986. It developed to epidemic proportions in the UK, reaching a peak in 1992. Although BSE is now fading away in the UK, cases have eventually turned up in many other countries.

2   Mainly through epidemiological studies, veterinary scientists quickly established (at least to their satisfaction) that BSE was caused by the inclusion in cattle feed of ruminant-derived MBM contaminated with a TSE-causing biological agent of some kind, following a change in the rendering process. The initial source of contamination may have been material from sheep carrying the TSE scrapie. Subsequently, increasing amounts of material from cattle carrying BSE would have become included in cattle feed.

3   Human TSEs include (classical) CJD (which may be sporadic, inherited or iatrogenic), GSS, kuru and vCJD.

4   vCJD was recognised in 1996 and linked to exposure to BSE in beef and beef products before 1989.

5   The number of suspected cases of human TSE diseases in the UK referred to the CJD Surveillance Unit rose from about 50 in 1990 to over 150 per year during 1997–2003. This increase may partly reflect greater awareness of TSEs among members of the medical professions.

6   Periodically, epidemiologists have predicted the eventual total number of vCJD cases in the UK in the form of a best estimate plus upper and lower 95% confidence limits. The upper 95% confidence limit, which was 10 million in 1997, had fallen to 50 000 by 2002. The 2003 prediction gave a best estimate of about 200 deaths by 2080, with upper and lower 95% confidence limits of 7000 and 10 respectively.

7   These predictions assume that vCJD is contracted only from having eaten contaminated beef (and not from infected surgical instruments or through blood transfusion). They also assume that only people with a particular genotype (*MM*) are susceptible to vCJD (which is now known not to be the case).

8   Analysis of tonsil samples is also being used to predict the likely total number of vCJD cases.

9   By the end of 2004, there had been 148 deaths in the UK from definite or probable vCJD.

10 In 1982, Stanley Prusiner first proposed that TSEs are caused by proteinaceous infectious particles or prions. These are protease-resistant particles (PrPs) that exist in (at least) two conformations – 'normal' PrP^C molecules (relatively rich in α-helices) and TSE-causing PrP^{Sc} molecules (relatively rich in β-sheets).

11 When PrP^{Sc} molecules interact with PrP^C molecules, the latter can be converted to PrP^{Sc}. Since PrP molecules from different species – with slightly different sequences of amino acids – can also interact in this way, TSEs can cross species barriers. Prusiner's protein-only hypothesis of TSEs is now very widely – but not universally – accepted.

12 There are some indications that prion-like behaviour might be quite widespread in nature and represents a previously unsuspected aspect of 'normal' biology.

Questions for Chapter 1

Question 1.2

(a) Identify and *carefully* explain any errors in the following statement: 'Prion diseases such as BSE and vCJD are caused by mutation of the *PrP* gene from the *PrP*^{C} allele to the *PrP*^{Sc} allele.'

(b) Rewrite this statement correctly.

Question 1.3

In prion diseases, the production of disease-causing PrP^{Sc} protein in a cell has been described as a 'chain reaction'. (a) Identify *two* ways in which the PrP^{Sc} molecule that initiates the 'chain reaction' could have arrived in the cell. (b) How are further PrP^{Sc} molecules produced in the cell? (c) How does the production of PrP^{Sc} molecules give rise to the symptoms of prion diseases such as BSE and vCJD?

Question 1.4

(a) In 2003, epidemiologists at Imperial College London estimated the likely number of deaths from vCJD in the UK by the year 2080. What was their prediction? (b) Suppose that the number of deaths from vCJD in the UK by 2080 turned out to be (say) 3000. To what extent would this invalidate the prediction? (c) Critically discuss the possible reasons for the discrepancy between prediction and eventual outcome.

Chapter 2

Managing the BSE/vCJD episode

Having concentrated in Chapter 1 on the 'science' behind BSE and vCJD, we now turn our attention to how the episode was managed by scientists, politicians and other relevant decision makers. Not surprisingly, we shall find that the themes of communication, risk and ethical issues are inextricably linked to that of decision making (at local, national and international levels).

Over the years, the UK Government implemented a great many Orders and Regulations, amending several of these more than once. The European Commission (EC) of initially the European Economic Community (EEC) and later the European Union (EU) also issued various Decisions, Directives, etc. Fully cataloguing the legal framework under which BSE (and hence vCJD) has been dealt with would be tedious in the extreme! Therefore, Sections 2.1–2.3 respectively present in outline a 'history' of how the episode was handled during the following periods:

- up to May 1990 (when an incident occurred that came to epitomise official attempts to reassure the public);
- from May 1990 to March 1996 (when the probable link between BSE and vCJD was announced);
- from March 1996 to the time of writing (2005).

Section 2.4 looks briefly at the official inquiries into BSE/vCJD. The international dimension is considered in Section 2.5.

Activity 2.1

Allow 15 minutes

It is rather easy to get bogged down in the minutia of the various 'story lines' running through an event of this complexity and significance. A 'BSE/vCJD Timeline' is therefore provided on DVD–ROM. As well as helping you 'see the wood for the trees', the Timeline will be updated periodically to reflect developments that occur after the time of writing (2005).

The resources and instructions for this activity have been placed on the DVD–ROM. To begin the activity you should now access the DVD–ROM through your computer. Once the S250 DVD–ROM has loaded, select **BSE/vCJD Timeline** and then enter the timeline. For this activity you should select the link called **Activity 2.1** and follow the on-screen instructions. Comments on this activity are provided with these on-screen instructions.

2.1 Up to May 1990

BSE was formally recognised as a new disease in November 1986. However, this information was kept under 'embargo' at first while an initial epidemiological study – involving the collection of data from 200 herds – was started. The Ministry of Agriculture, Fisheries and Food (MAFF) was officially informed about BSE by the Chief Veterinary Officer (CVO) in June 1987. By December

C D
Initially not made public.

1987, those responsible for analysing the data from the initial epidemiological study had concluded that the only viable hypothesis for the cause of BSE was contamination of MBM derived from ruminant animals. In early 1988, officials started to check with individual producers of cattle concentrates what had been included in the MBM they had used; their responses provided further confirmation of the MBM hypothesis. From June 1988, BSE was made a notifiable disease (i.e. a disease that by law must be reported to the authorities) and cows suspected of having BSE had to be isolated when calving. From July, the sale, supply and use of feed for ruminants that contained protein (except milk) derived from ruminants were prohibited until the end of 1988. This is known as the **ruminant feed ban**.

concluded that BSE originated from contaminated feed.

R D

■ Do you think this was an adequate initial response to BSE?

■ Given the relatively novel nature of BSE – and refraining from hindsight – it is difficult to see what further precautionary measures ought to have been introduced at this stage to reduce risk. Within about 12 months of MAFF being officially informed of an entirely new disease, arrangements had been made that should have ensured all cases of BSE were recorded and BSE's most likely transmission route to other cattle closed off (albeit temporarily, in the first instance). Even the delay of about six months between BSE being recognised as a new disease and government ministers being informed is understandable given that officials needed time to assess the magnitude of the problem and politicians and officials usually have more pressing matters to consider than a new disease in cattle that at the time did not seem to pose any risks to human health.

not considered any threat to humans.

Cases of BSE in the UK continued to rise (Figure 1.1) and it was therefore announced in April 1988 that a Working Party, chaired by Professor Richard Southwood of Oxford University, would investigate BSE. The Southwood Working Party, which held its first meeting in June 1988, welcomed the ruminant feed ban and also recommended that affected cattle should be destroyed. In August 1988, Orders came into effect implementing the recommended slaughter policy and authorising compensation to be paid at 50% for confirmed cases of BSE and 100% for slaughtered cattle that turned out not to have BSE.

E D

■ What were the likely consequences of such differential compensation?

■ Because farmers would be given only 50% compensation for confirmed cases of BSE, it would not be surprising if some BSE-affected cattle were passed off as healthy and entered the human food chain before their symptoms became obvious to those not familiar with the temperaments of individual animals.

Because of this possibility, full compensation for affected animals was eventually paid from February 1990.

R E D

Particularly after confirmation in March 1996 that BSE was probably the cause of vCJD (Section 1.1.4), differences in the early days of BSE between Southwood on the one hand and officials and politicians on the other started to emerge. For instance, Southwood claimed that he advised on full compensation right from the start but that MAFF would not allow this in order to save money.

However, officials and politicians maintained that the recommendations of scientists were always implemented; but how quickly and willingly? Was some of Southwood's advice influenced by what he thought would be politically acceptable? It must be borne in mind that there was considerable uncertainty at this time as to the magnitude and seriousness of the BSE problem. Moreover, all concerned would have been acutely aware of the possible adverse economic consequences of their decision making for large numbers of people dependent on the agriculture and food industries. This highlights some of the challenges of taking precautionary measures. Decision makers often have to make judgements in the light of competing factors.

various factors affect decisions

In November 1988, as a further precaution, the Southwood Working Party advised that milk from infected cattle should be destroyed. It also recommended extension of the ruminant feed ban. This was first extended to the end of 1989 and then the time limitation was removed completely.

D

In February 1989, both the Southwood Report and the government's response to it were published. The government accepted all the recommendations in the report, including establishment of the Tyrrell Committee on scientific research into BSE. Ministers received the Tyrrell Report in June 1989 and it was published – together with the government's response to it – in January 1990. All top and medium priority research projects recommended by Tyrrell were approved. It was emphasised at the time that the six-month delay in publication was related to making arrangements for the various projects to be properly funded and that the research itself had not been delayed. BSE (and TSEs more generally) thus became a major area for new research funding.

C D

Reports published. government acted on all recommendations. funding approved for further research.

In June 1989, it was announced that specified bovine offals (SBO) were to be banned from all human food. This decision was implemented in November 1989 for England and Wales and in January 1990 for Scotland and Northern Ireland (whose legal systems are independent of that of England and Wales). This is interesting because Southwood had recommended only that SBO should be excluded from baby food. This is an example of the precautionary principle in operation – decision making that goes beyond the current scientific knowledge of risk. It is important to note that when the Secretary of State for Health made his initial statement about vCJD in the House of Commons in March 1996 (Section 1.1.4), he said that victims were probably exposed to BSE prior to the introduction of the SBO ban.

R D

scientists recommended banning SBO only in baby food over-ruled by government and banned in all human food.

D

So far we have considered only local and national decision making: decision making at the international level also had a role to play. For example, during March and April 1990, the EC restricted export of cattle from the UK to animals that were less than six months old when slaughtered, ruled that all cases of BSE had to be notified to the Commission and banned the export of SBO-containing material from the UK to other Member States. Thus, at this stage the EC appeared to be applying the precautionary principle more comprehensively than did the UK government.

Meanwhile in the UK, it was announced in April 1990 that the Spongiform Encephalopathy Advisory Committee (SEAC) would be formed – effectively, a (semi-)permanent advisory body replacing the Southwood Working Group and with a similar remit. This suggests acceptance that BSE was going to be a long-term issue.

CRED

In May 1990, the then Secretary of State for Agriculture, Fisheries and Food (John Selwyn Gummer) and the CMO made the first of several declarations that beef was safe to eat, an example of which is illustrated in Figure 2.1. In the following activity, you will investigate the impact of statements such as these by considering the four themes: communication, risk, ethical issues and decision making.

Figure 2.1 The then Secretary of State for Agriculture, Fisheries and Food, John Selwyn Gummer, and his daughter, Cordelia, eating beefburgers. (First broadcast on BBC2, Newsnight, 16 May 1990.)

Activity 2.2

Allow 45 minutes

Now return to the **BSE/vCJD Timeline** on the DVD-ROM, select the link to **Activity 2.2** and follow the on-screen instructions. Comments on this activity are provided with these on-screen instructions.

2.2 From May 1990 to March 1996

The discovery of FSE in the domestic cat and TSEs in antelopes of five different species (Section 1.1.3) – plus laboratory transmission of BSE to a pig – confirmed the transferability of BSE between species. The ban on SBO was therefore extended from September 1990 to cover *all* animal feed (including pet food). At the same time, the export of such feed to other Member States of the EU was banned. Nevertheless, the Tyrrell Committee advised that there were no implications for human health. An October 1990 Order brought in new arrangements whereby cattle farmers had to keep records of their animals for 10 years.

BSE transmissable between species, yet Tyrrell committee advised no danger to humans.

■ Bearing in mind the apparently inexorable growth in the number of BSE cases at this time (Figure 1.1) and the appearance of TSEs in a range of other species, do you think that an even more precautionary approach would have been justified? Are any 'mixed messages' evident in how the episode was being handled?

■ There are contradictions in the introduction of a range of control measures designed to stop the spread of BSE among cattle and to other species, while at the same time insisting that beef was entirely safe for human consumption. There could have been no scientific evidence that BSE posed no health threats to humans. Indeed, there could never be such evidence because it is logically impossible to prove a 'negative'. Strict application of the precautionary principle would therefore suggest that more action ought to have been taken even at this stage. On the other hand, the beef and dairy industries were important for the UK economy and this also had to be borne in mind.

Notwithstanding the various precautions outlined above, it was announced in November 1990 that BSE had been detected in offspring born after the ruminant feed ban was introduced in 1988.

■ What are the implications of this announcement?

■ Despite the precautionary control measures that had been introduced up to this point, BSE clearly was still being contracted even by calves that should not have consumed any SBO (and hence PrP^{Sc}) in their feed. Either BSE was being transmitted through routes other than contaminated SBO in cattle concentrates or the controls were not operating effectively.

Clearly, it was essential to establish whether the scientific understanding of BSE was incomplete or whether human failings meant that the various precautionary measures that had been introduced were not having the desired effect.

The controls certainly did not always operate as intended. First, a ban might simply be ignored. For instance, if you were running a dairy farm and found that you had some cattle concentrate containing SBO left after July 1988, might you not be tempted to use it – either to avoid waste or because of commercial pressures you were under? Second, accidents happen. If SBO is banned from ruminant feed, but not from feed intended for other animals, there can be genuinely accidental contamination either during manufacture of ruminant feed or on a farm. (Of course, if tissues from the non-ruminant animals to which SBO-containing feed was given were then incorporated into cattle feed, then the cycle of contamination would continue.)

■ With SBO banned from ruminant feed in the UK, how would feed manufacturers probably respond?

● They would probably still continue to export SBO-containing material provided there was a market for their product. Indeed, they might endeavour to expand this market in an effort to compensate for the loss of their home market. At the time, this would have been perfectly legal even if ethically questionable, given that their product was not considered sufficiently safe for use in the UK.

R D

beef and dairy important to UK economy.

2 years after feed ban BSE still found in calves.

R E D

ban could be ignored or it could be feed accidentally. Ban on SBO was extended to all animal food only from 1990

R E D

D

Ban on export of SBO to all countries as animal feed could still export as fertiliser.

R D

R D

first indication that there may be concerns regarding the transfer of BSE to humans.

C R D

C R D

C R D

C

A number of regulations were introduced to reduce the risk to human health.

R D

Even after the EC banned export of SBO to Member States, manufacturers continued to export feed containing SBO to countries outside the EU. However, the Department of Trade and Industry introduced an Order in July 1991 controlling the export of SBO and feed containing SBO to such countries.

In November 1991, it became illegal to use MBM produced from SBO as an agricultural fertiliser. This was designed to address the problem of back-importation – SBO legally exported to continental Europe for use as a fertiliser and then legally re-imported for feed because of the shortage of feed in the UK.

In March 1992, the use of cattle heads once the skull had been opened and the brain removed was prohibited except in areas that are free at all times from any food intended for human consumption. This was a recommendation from SEAC, which concluded that the measures then in place provided adequate safeguards for both human and animal health. Given the mounting evidence that previous safeguards had not proved adequate, the basis for this reassurance is unclear. Certainly, there appeared to be no new scientific evidence to support it.

In March 1993, the CMO again made a public statement declaring that beef was safe to eat.

Over the next three years, a number of new and amended Orders and Regulations were introduced in the UK, as well as rules that applied throughout the EU. These were all designed to bring the UK epidemic under control, prevent BSE spreading elsewhere, safeguard human health and reassure the public. For example, mammalian protein was prohibited from being fed to ruminants throughout the EU; control of bovine offal was extended to cover the animals' thymus and intestines; TSEs in all species – not just ruminants – were made notifiable diseases; and the spinal cord (plus obvious nervous and lymphatic tissue) had to be removed from bovines over six months old and not used for human consumption.

In November 1995, MAFF officially informed SEAC that some abattoirs were ignoring the ban on SBO.

■ A 1995 Regulation required that SBO be stained blue (Figure 2.2). What does this requirement, which effectively removed decision making from the local level, suggest?

▪ Without such staining to make it obvious, SBO might be getting into food for human consumption. The blue coloration served to warn workers in abattoirs that they were dealing with banned materials and also that processing it would be pointless as nobody would choose to eat blue food.

At about this time, SEAC turned its attention to the possible dangers posed by the removal of so-called **mechanically recovered meat (MRM)** from the spinal column. Special equipment was used to obtain almost every last scrap of usable meat from a carcase for inclusion in cheaper meat products. In the process, there was a possibility that small amounts of nervous tissue contaminated with PrP^{Sc} might end up in food for human consumption and thus pose a risk of transferring BSE to humans. The use of bovine vertebral columns in the manufacture of MRM, the use of MRM in food for humans and the export of bovine MRM to other EU countries were all prohibited by a December 1995 Order.

Figure 2.2 Blue-stained SBO (specified bovine offals).

2.3 From March 1996

C

In March 1996, SEAC announced that the CJD Surveillance Unit had identified vCJD as a new human disease, the first death from which occurred in May 1995. SEAC concluded that, although there was no direct evidence of a link, the most likely explanation for vCJD was exposure to BSE before the SBO ban was introduced in 1989. At the time, the strongest evidence for the link was that vCJD was a *new* TSE in humans (the symptoms of which differed from previously known human TSEs) that had arisen about a decade after BSE, a *new* TSE in cattle. The link was confirmed only later through similarities between the conformation of the PrP^{Sc} molecules in humans with vCJD and in cattle with BSE (Chapter 1).

C R D

The Secretary of State for Health made a formal statement about the likely link between BSE and vCJD in the House of Commons on 20 March 1996. Confirmation that there was almost certainly a link between BSE and vCJD ('the human form of mad cow disease') – despite repeated assurances over the years from politicians, officials such as the CMO and public bodies such as SEAC – represents a dramatic turning point in the BSE story. Of course, there was extensive media coverage of this development (we will look in detail at a contemporary newspaper article in Chapter 3). Public confidence in the safety of British beef was severely dented and sales of beef and beef products fell dramatically. The price of beef products in the shops also fell, as attempts were made to stabilise the market.

R E

■ Was it right to encourage people to buy food that *might* not have been entirely safe by offering it at bargain prices?

By 1996, after a whole series of precautionary measures had been introduced to safeguard public health, managers in the retail trade were presumably completely convinced that beef was perfectly safe to eat. On the other hand, they would also be acutely aware that many people's livelihoods (including their own to some extent) would be jeopardised if the beef industry were allowed to collapse.

R D

Government
continue to take
precautionary
measures .

At the same time, the government also announced a further precautionary measure – again following advice from SEAC – that cattle over 30 months old had to be deboned at specially licensed plants and the trimmings kept out of the food chain (this was known as the **30+ months ban**). Furthermore, mammalian MBM was to be banned in *all* feed for farm animals (i.e. not only feed intended for ruminants or even mammals).

■ What do these two measures suggest?

▨ The first suggests that the authorities considered that cattle older than 30 months might pose a greater risk to humans than younger cattle. The second suggests that the authorities suspected that feed intended for non-ruminant farm animals (e.g. chickens) might being eaten by cattle and other ruminants.

D

Beef export ban -
serious threat to
U.K. economy .

Also in March 1996, the EC prohibited export from the UK of live bovine animals; bovine semen and embryos; the meat of bovine animals slaughtered in the UK; products of bovine animals slaughtered in the UK liable to enter the food chain or destined for use in medicinal products, cosmetics or pharmaceutical products; and mammalian-derived MBM. In short, the EU banned all British beef exports (the **beef export ban**). A major priority of the UK government over the next few years would be to convince fellow members of the EU that sufficient measures had been taken to ensure that infected materials could not enter the human food chain. The success or otherwise of this campaign would, of course, be judged by the lifting or continuation of the EU ban.

D

Not surprisingly, there followed a whole series of new and amended Orders and Regulations in the UK, most of which were designed to tighten up existing controls. Several of these were related to the decision that animals over 30 months old at the time of slaughter should not be allowed to enter the human food or animal feed chains.

R

■ What would be the scientific rationale for this?

▨ It was believed that most animals that contracted BSE did so as a result of consuming contaminated feed as calves. Because SBO had already been officially banned from cattle feed for many years by the mid-1990s, the probability of vCJD being transmitted to humans or BSE to other cattle should by then have been very low. Like all TSEs, BSE has a moderately long incubation period (typically 3–5 years) and so infected cattle were thought not to contain much PrP^{Sc} until they were over three years old. By ensuring that no animals older than 30 months entered the human food or animal feed chains, the ban should have reduced the risk even further.

A politically astute aspect of the 30+ months ban is that 'best beef' is eaten at less than 30 months old anyway. The ban was therefore designed to have the minimum possible adverse effect on the 'best beef' industry. Remember that BSE was largely a problem of dairy cows that were usually sent for slaughter when their milk yields declined after the age of five years.

Ministers announced in December 1996 that the backlog of animals that had to be slaughtered and disposed of under the 30+ months ban had been cleared.

■ In the latter half of 1996, the Feed Recall Scheme was launched with the aim of collecting and disposing of any MBM and feed containing MBM present on farms, in feed mills or at feed merchants. To what extent do you think that this was appropriate?

▨ Given that it had long been believed that BSE had been caused by contaminated feed and that it was known that some cattle born after the introduction of the SBO ban had developed BSE, recalling potentially contaminated feed was an appropriate course of action. However, it does seem rather surprising that such a scheme had not been organised much earlier. Presumably, the emergence of vCJD had convinced officials that this additional precaution was now urgently necessary. In effect, this measure removed decision making from the local level – farmers cannot accidentally or deliberately use MBM-containing feed if they don't have access to it.

R D

Feed recall scheme to reduce risk of contaminated feed still in food chain.

The system of record keeping in the beef and dairy industries was enhanced considerably. For instance, from January 1999 one ear tag had to be permanently attached to a calf within 36 hours of birth and a second within 30 days (see cover picture). All cattle born from July 1996 also had to have mandatory cattle movement documents (so-called 'cattle passports'). Northern Ireland, which is particularly reliant on its beef exports, introduced at a relatively early stage a sophisticated and comprehensive computerised system for tracing animals. The existence of this system convinced the EC to lift the ban on beef exports from Northern Ireland more than a year before it decided to lift the ban on exports from the rest of the UK.

C D

ear tags & cattle passports allow tracking of animals

■ What were the benefits of introducing such enhanced record keeping?

▨ First, it should help ensure that cattle older than 30 months did not enter the human food chain. Secondly, when an animal was discovered to have BSE, possible sources of infection on the farm on which it was raised could be investigated and other animals that might have been infected at the same time could be traced for examination.

C R D

In May 1996, within two months of the announcement on vCJD, the UK government sent to the EC details of its BSE Eradication Programme, which was approved in principle the following month. At about the same time, the European Council meeting in Florence agreed a framework for lifting the ban on beef exports from the UK. This framework specified five preconditions that the UK was required to meet. One of these was implementation of a selective cull of cattle deemed to be most at risk of developing BSE. The UK government announced details of this selective cull in August 1996. However, following publication in *Nature* of the results of a major epidemiological survey by a team lead by Professor Roy Anderson of Imperial College London which suggested that the BSE epidemic would virtually die out around 2001 *even if no further measures were taken*, the UK government announced a month later that it would

government questioned the strategy for the selective cull so there was a delay in it being carried out.

not be proceeding immediately with the cull. The reason given for this decision was that further scientific research needed to be done to establish the most appropriate culling strategy in the light of both the recent epidemiological study and preliminary results from MAFF (confirmed by SEAC in April 1997) which suggested some degree of cow-to-calf transmission of BSE. Such cow-to-calf transmission might represent genetic susceptibility to acquire infection from feed or transmission of the infectious agent through either the placenta or milk. In fact, the data suggest (but do not prove) the genetic explanation. Nevertheless, it was announced in December 1996 that the selective cull would go ahead after all and an Order to this effect came into force in January 1997 (along with another that dealt with the compensation to be paid for animals slaughtered under the cull).

R D

■ Do these developments suggest there was a strong scientific basis for the cull?

■ No. It appears that the selective cull was mainly a precautionary measure designed to reassure the rest of the EU as part of the UK government's continuing effort to get the beef export ban lifted.

C

In parallel with the cull, the Date-Based Export Scheme (DBES) was developed in the hope that the ban on beef exports would be lifted from herds that were certified as having been BSE-free for a certain period of time.

The General Election in May 1997 brought Labour to power in the UK, with a huge majority in the House of Commons, after 18 years of rule by the Conservatives. The new administration continued to work towards persuading the EU to lift its ban on beef exports.

C R E D

Following a review conducted by MAFF and the Department of Health, in September 1997 the UK government confirmed that SEAC still had a key role to play in advising on BSE and vCJD. Shortly after announcing in October 1997 that it believed no further measures governing beef or beef products for human consumption were necessary, SEAC recommended that all beef derived from home-produced or imported cattle over six months old at slaughter should be deboned before sale to consumers. So-called 'beef on the bone' was banned from December 1997. The issue of beef on the bone then became something of a *cause célèbre* through 1998 and 1999 as politicians (mainly from opposition parties) and others took advantage of 'photo opportunities' to deliberately flout the ban in order to demonstrate their conviction that 'British beef was safe to eat'. They were certainly not taking an approach based on the precautionary principle. Rather, they challenged official decision makers through publicity stunts – a form of 'direct action'. Although the CMO advised in February 1999 that the ban should continue, it was finally lifted in December of that year.

publicity stunts by politicians and others following "beef on the bone ban"

C R D

The beef export ban was far more significant economically than the issue of 'beef on the bone'. In May 1998, the European Court of Justice upheld the validity of the ban on UK beef. However, the export ban was lifted for Export Certified Herds Scheme beef from Northern Ireland in June 1998. Then, from August 1999, the export ban was lifted for DBES beef from the whole of the UK. The French Food Standards Agency immediately expressed its concerns about the safety of British beef and France declined to lift the ban. The EC's Scientific Steering Committee unanimously concluded that it did not share these concerns about beef and beef products exported under DBES. In November 1999, the UK formally asked the Commission to take action against France for refusing to lift its ban and later that month the Commission asked the French government to review its decision. During December, the French government stated its intention to retain the ban. The Commission then issued a Reasoned Opinion (a legal device) on France's failure to lift the ban, which France responded to by insisting that it would maintain the ban. Finally, the Commission announced that it would pursue the case through the European Court of Justice.

French refused to lift ban!

D

While this disagreement between France and the Commission (as well as the UK) was developing, there was some concern that Germany would also maintain a ban on importing British beef. However, the German government insisted that any delay was due to a constitutional requirement to consult the constituent Länder (or states) that make up the Federal Republic before such a change of policy could be made. Indeed, the Bundesrat (the second chamber of the Federal Republic, in which the Länder are represented) voted in favour of lifting the ban in March 2000 and it was formally lifted later that month.

Germany lifted ban in March 2000

C R D

It took until December 2001 for the European Court of Justice to rule that, by refusing to permit the marketing in its territory after 30 December 1999 of DBES beef which had been correctly marked or labelled (and which should have served to devolve decision making about risk to the consumer), France had failed to fulfil its obligations. During early 2002, the EC employed various devices to ask formally for an explanation of France's failure to comply with the European Court of Justice's ruling. This culminated in June 2002 in the Commission issuing a further Reasoned Opinion to the French government and giving it 15 days to respond. In July, the Commission requested that the Court impose a penalty of €158 250 (about £110 000) per day on France for non-compliance with the Court ruling that its ban on the import of UK DBES beef was illegal. The French Food Standards Agency announced in September 2002 that British beef no longer posed a risk to French consumers and the French government formally lifted the ban the following month. In response, the EC announced in November that it was dropping the European Court of Justice case to impose financial penalties on France for illegally banning British beef. For the first time since March 1996, British beef could again be marketed throughout the EU and elsewhere.

France lifted ban in October 2002 - finally!

C R D

■ What do you think motivated the French government to maintain its ban for so long?

It could be argued that it was simply following independent science-based advice from its national Food Standards Agency. In other words, it was applying the precautionary principle. However, it is unlikely that the French government was unaware of the potential commercial advantages to France's own beef industry of continuing the ban for as long as possible. The battle between France and the EC (and, by implication, the UK) certainly engendered a great deal of adverse publicity for British beef. Of course, the eventual removal of the ban did not ensure that French consumers would actually buy British beef! (This is an example of decision making at a local level.) Ironically, in this instance more informative labelling of the origins of a product may have worked against the export of British beef to France.

On the other hand, the effects of BSE on British farming – and eventually on the health of some UK residents – were so severe that it is hardly surprising that a neighbouring country had serious reservations about allowing resumption of British beef imports.

D

In December 2004, the UK government announced that the 30+ months ban would be phased out over the succeeding months.

2.4 The official inquiries

C D

In December 1997, the UK government announced the Public Inquiry into BSE that was to be chaired by Lord Justice Phillips. Public hearings commenced in March 1998 and eventually concluded in December 1999. The Report of the BSE Inquiry was delivered to the Secretaries of State for Agriculture, Fisheries and Food and for Health and published in October 2000. Having published an interim response in February 2001, the government's final response to the report was published in September 2001. As we saw in Chapter 1, the BSE Inquiry concluded that the disease probably originated in cattle in South-West England during the 1970s or early 1980s. A Committee chaired by Professor Gabriel Horn was then asked to carry out a rapid review specifically into the origins of BSE. The Horn Committee published its report in July 2001, concluding that scrapie could not be ruled out as the source of BSE. We will return to the BSE Inquiry in Chapter 3.

2.5 The international dimension

Although by 2000 the number of new confirmed cases of BSE in the UK was approaching zero (Figure 1.1), several of these involved cattle born since the tightening of feed controls in August 1996. By then, confirmed cases of BSE were also being officially notified in countries other than the UK (e.g. in cattle born in France, Spain, Germany and Japan; see Figure 1.2). Indeed, in September 2001 it was predicted that during 2002 there would be more cases of BSE in France than in the UK.

BSE confirmed in cattle in other countries.

■ What is the likely cause of these BSE cases in animals born in continental Europe or elsewhere?

▨ As calves, these animals may well have been fed contaminated cattle concentrates exported from the UK before effective controls were introduced in the early 1990s.

R E

The UK exported 25 000 tonnes of MBM in 1991. By then, MBM destined for Member States of the EU did not include SBO. However, MBM sent outside Europe (e.g. to Thailand, Taiwan, Singapore and Indonesia) did contain SBO. Some of this MBM was subsequently sold on to other countries, such as China. When the UK stopped exporting MBM entirely in 1996, other European countries took over the trade – until 2000, when BSE cases were reported from all over Europe. Then the USA took it over, having declared for years that it was entirely free of BSE. This emphasises the truly globalised trade in animal feed – as well as animals destined for human consumption – the origins of which can be extremely difficult to trace.

In fact, EC scientists predicted in 2000 that BSE might already exist in the USA, because prior to 1996 it had imported British cattle (some of which *could* have had BSE) and 44 tonnes of British MBM. After 1996, the USA imported 800 cattle from other European countries in which BSE was subsequently discovered.

R D

In May 2003, a Black Angus beef cow born in Canada in 1995 was found to have BSE. Canada's only previous case (in 1993) had been born in Britain in 1987. The USA immediately closed its borders to Canadian cattle. However, in 2002 Canada had sent 500 000 live cattle and a great deal of MBM to the USA.

R D

After behaving strangely at a slaughterhouse in Washington State in December 2003, a six-year-old Holstein cow was confirmed to have BSE. The cow had almost certainly been infected from cattle feed manufactured in either Canada (where it had been born) or the USA that had included tissue from native-born animals. Confidence in the USA's beef industry was severely damaged. For instance, countries such as Japan (a major purchaser of US beef) – and, indeed, Canada – refused to import beef from the USA. On the basis of BSE in one animal, this might be regarded as an extreme example of the application of the precautionary principle. The USA commenced an enhanced surveillance programme in June 2004 using a rapid test for BSE. This produced two inconclusive results in June/July 2004 and another in November 2004. All three animals tested negative in a confirmatory test based on immunohistochemistry (which detects small quantities of substances based on their binding to specific antibodies). Although the first two animals also tested negative in a further Western blot test (another very sensitive immunohistochemical technique), the November 2004 case came back positive. Meanwhile, Canada confirmed two more BSE cases in January 2005, bringing to four the total number of BSE-infected cows identified or linked to Canada.

BSE discovered in U.S.A.

By 2003, diagnostic tests were available to test for BSE in live cattle. This represents an advance on the way BSE cases were recognised in the UK in the 1980s and 1990s, when confirmation of suspected cases required post-mortem examination of brain tissue. However, there were challenges to the science from within the beef industry, not least because there was no consensus on which of several tests was the most effective in identifying BSE or how many – and which – cattle needed to be tested in order to establish the incidence rate of BSE in the North American herd.

R E D

■ In 1997, the USA banned the feeding of cattle carcases to other cattle – although it was widely believed that this ban was not enforced properly (a situation reminiscent of that which applied in the UK in the early years of BSE). However, it remained legal to include cattle MBM in feed for pigs and poultry. It was also legal for pigs and poultry – and poultry litter containing poultry feed – to be included in cattle feed. Explain why, in these circumstances, the ruminant feed ban was extremely unlikely to have been adequate to stop the spread of BSE.

▨ For a start, accidental cross-contamination could occur during the manufacture of feed for cattle, pigs and poultry. It is also possible for cattle accidentally or deliberately to be given feed intended for pigs or poultry. Moreover, cattle MBM could legally be incorporated into feed given to pigs or poultry and then tissues from *these* animals – and also some of their feed – to be incorporated into cattle feed. If even tiny amounts of PrP^{Sc} were contained in the cattle MBM, then BSE could spread.

Dealing with vCJD

Although the UK's BSE epidemic had been brought under control by 2004, relatively small numbers of cases of BSE continue to arise throughout the world. It would now take extreme carelessness for a BSE outbreak on the scale of that experienced in the UK to occur in another country. The concern now is that *any* level of BSE in a beef herd might cause some cases of vCJD. By the end of 2004, there had been 148 deaths from definite or probable vCJD in the UK. There have also been a small number of cases elsewhere (e.g. one in the USA in 2002 involving a woman who had been brought up in the UK, another in Ireland in 2004 involving a man who had never lived in the UK, and several in continental Western Europe).

C R E D

As already discussed in Chapter 1, precautions have been put in place to guard against the possibility of vCJD being passed on in blood transfusions or through the use of contaminated surgical instruments. There is also ongoing research into possible treatments for vCJD. Although drugs intended for treating diseases in humans normally take *many* years to develop because of the rigorous testing for efficacy and safety insisted upon by regulatory authorities and the requirement that informed consent be given by participants in trials, some exceptions have been made in the case of vCJD patients. For instance, in 2001 Stanley Prusiner was given permission to treat a UK patient with the antimalarial drug quinacrine. Unfortunately, this patient eventually died, apparently from liver complications triggered by the drug. However, Prusiner's team is trying to improve the drug's efficiency – for instance, by fusing together two quinacrine molecules. Pentosan polysulphate (PPS) is another drug developed for other purposes (in this case, treating infections of the urinary tract) that has been used to treat vCJD. PPS appears to stop abnormal prion proteins from forming clumps that cause neurons to die. One patient, who had been given only hours to live before PPS treatment began in 2003, was still alive 18 months later. Although PPS may have slowed down or even stopped vCJD's progress in this patient, it is certainly not a cure for the disease.

still working on a cure for vCJD.

R E D

- Is it right to 'fast track' possible vCJD treatments in this way?

- So far, vCJD has invariably proved fatal for the often young patients suffering a debilitating illness and their families are often desperate. It could therefore be argued that the purpose of these treatments, treating fatally ill patients, is ethically justified even though the process falls outside conventional ethical procedures.

On the other hand, this argument is seldom employed successfully for other diseases. Proper clinical assessment of new treatments almost always involves taking elaborate precautions against bias – such as adequate replication and randomised double-blind trials (in which neither the patients nor those administering the treatments or assessing their effects know which patients received the treatment under test and which a supposedly non-effective placebo). Such objective assessment is hardly possible for vCJD, particularly when drugs are used on an *ad hoc* basis on individual patients at various stages in the development of the disease. This makes it extremely difficult to decide whether a treatment has 'worked' and hence whether it would be effective if used for other patients.

treatments used without proper clinical trials make it difficult to prove their effectiveness.

Another line of research is to test tissue from tonsils removed during routine tonsillectomy operations for the presence of PrP^{Sc} protein in order to assess the prevalence of vCJD the general population (Chapter 1).

CRED

■ In this type of research, it is usually arranged that the individuals from whom the tonsils were removed cannot be identified. What issues does this anonymity suggest to you?

▨ In the absence of a cure for vCJD, should individuals be told that they *might* develop vCJD at some stage in the future? If so, how should they be told? What other information and support should be provided? If someone was told that they were incubating vCJD, this knowledge would almost certainly have major effects on their lives – ranging from possible psychological damage to life insurance implications (see discussion of risk and actuarial data in the *Introduction to the course*). The presence of PrP^{Sc} protein might not lead inevitably to the development of vCJD. Even if it did, the likely timescale might well mean that many people would die of other causes before the symptoms of vCJD manifested themselves. On the other hand, if a treatment were developed that could control the development of vCJD if given early enough, then the question arises whether 'at risk' members of the population should be identified and given the option of availing themselves of the treatment.

should we test for PrP^{Sc} protein in tonsil tissue?

Doubtless, the BSE and vCJD 'stories' will continue to develop during the years in which this course is presented. Moreover, it is entirely possible that TSEs will continue to challenge us to refine our ideas about some aspects of biology. Further issues related to the course themes are also bound to arise.

In Chapter 3, we will look at aspects of the legacy of the BSE/vCJD episode which continues to influence the relationship between science and wider society – for instance with regard to public attitudes to genetic manipulation and nanotechnology.

Activity 2.3

Allow 15 minutes

Now return to the **BSE/vCJD Timeline** on the DVD-ROM, select the link to **Activity 2.3** and follow the on-screen instructions. Comments on this activity are provided with these on-screen instructions.

Summary of Chapter 2

1 Once BSE had been identified as a new disease in cattle and veterinary scientists had become convinced that it was caused and spread by the inclusion of contaminated ruminant MBM in cattle concentrates, a series of decisions were taken at local, national and international level with the aim of bringing BSE under control and safeguarding human health (risk and decision making).

2 At least initially, these measures proved inadequate to stem the BSE epidemic. This was partly due to the controls often being flouted, either accidentally or deliberately (risk and ethical issues).

3   Nevertheless, there were repeated assurances from politicians and officials that British beef posed no health risk to the public (communication, risk, ethical issues and decision making).

4   When the likely link between BSE and vCJD was recognised in 1996, further precautionary measures were introduced in the UK and the EC imposed a ban on the export of British beef. Over several years, the UK government expended considerable effort to get this ban lifted. Although the ban was finally abolished in 1999, the French government refused to lift its own ban until 2002.

5   There have now been cases of BSE in many other countries, including the USA and Canada.

6   Since BSE is unlikely to flare up again as a major disease in cattle anywhere in the world, the emphasis now is on ensuring that even a low rate of BSE does not lead to cases of vCJD.

7   Efforts are also being made to ensure that vCJD is not spread through blood transfusions or the use of contaminated surgical instruments, and to find treatments for vCJD.

8   Those responsible for managing the BSE/vCJD episode often had to take decisions that affected people's lives and livelihoods on the basis of incomplete and sometimes contradictory scientific data and understanding (risk and decision making).

Questions for Chapter 2

Question 2.1

(a) What was the ruminant feed ban and when was it introduced? (b) What was the scientific basis of the ban?

Question 2.2

(a) What evidence is there that the ruminant feed ban did not work as effectively as it should have? (b) In what way(s) did the ban not work as intended?

Question 2.3

To what extent was banning cattle SBO and staining it blue an example of the precautionary principle in practice?

Question 2.4

(a) What cures for vCJD had been developed at the time of writing (early 2005)? (b) What issues are raised by attempts to find a cure for vCJD?

The legacy of the BSE/vCJD episode

The previous two chapters outlined the biology of prion diseases and the main
events in the management of the BSE/vCJD episode, a topic that has become
synonymous with issues of food safety, intensive farming practices, scientific
uncertainty and trust in scientific expertise. This chapter considers the legacy of
the BSE/vCJD episode by examining the impact of this topic on the relationship
between the scientific community, politicians, farmers, the food industry and the
public. How is it possible to judge this legacy? There are many answers to this
question involving issues of considerable complexity and subtlety and it would be
impossible to do justice to all the nuances here. Instead, this chapter considers
three aspects of the legacy of the BSE/vCJD episode:

*   developments affecting the scientific community
*   evidence of change in the relationship between science and society
*   lessons learnt from the Public Inquiry that examined BSE and vCJD.

In addressing these issues, you will be asked to reflect on what you have learnt
about the biology of prion diseases and the themes from the *Introduction to the
course* and Chapters 1 and 2.

It is important to note, of course, that you are being asked to reflect on these
issues with the benefit of hindsight. Although considerable uncertainties remain,
with respect to vCJD in particular, there is a much clearer scientific
understanding of BSE/vCJD now than there was in the mid-1980s. You should
bear this in mind when considering the issues described in this chapter.

3.1 BSE/vCJD and the scientific community

First identified in cattle in the mid-1980s, BSE is an example of an emerging
disease that has become the object of a considerable amount of scientific
research. Investigations continue to address the significant uncertainties that
remain in relation to this new illness. This is also true of vCJD, as scientists
continue to investigate human TSEs. The first legacy of note in relation to BSE and
vCJD is that the scientific study of these diseases has grown, particularly with
respect to biological investigations of prion diseases. As a result, scientists know
considerably more about TSEs and prions than they did in 1984 and a new scientific
consensus is emerging around the idea that prions cause TSEs. This has happened,
in part, because funding for scientific studies into the biology of human TSEs has
also grown over the last 20 years. As you saw in the *Introduction to the course*,
contemporary scientific investigations are expensive; they require considerable
resources, in terms of expert scientists and technicians, and also equipment. Studies
of BSE/vCJD are no different in this respect. A second legacy of the BSE/vCJD
episode is therefore the creation of the 'scientific infrastructure' required to
investigate these diseases, in part because decision makers (e.g. policy makers in
government) have identified these issues as important areas of study. But how do
these policy makers know which issues are important to investigate and which
issues to prioritise in terms of managing risk?

CRD

Increase in scientific research has lead to more understanding about BSE, vCJD and prion diseases considered important ∴ given funding.

87

New bodies, such as SEAC and the CJD Surveillance Unit, have been introduced to collate new scientific evidence and provide guidance to policy makers. This guidance informs decisions about the management of risk, including the imposition and, more recently, the removal of precautionary measures. These bodies also provide useful guidance with respect to informing the public about these developments, either directly, for instance through websites, or via third parties such as media professionals. Moreover, these new bodies have a remit to improve the co-ordination of research, strategically targeting important areas for investigation. A further legacy therefore involves changes to scientific practices in terms of the themes of communication, risk and decision making.

3.2 BSE/vCJD and the relationship between science and society

From the first official recognition that BSE existed in the UK cattle herd, through to the March 1996 announcement in Parliament and onwards to current debates, scientific research, consumer confidence, public officials and the beef industry have been inextricably linked through the themes of communication, risk, ethical issues and decision making. This makes the BSE/vCJD episode a particularly useful example for investigating science in context. This was a developing scientific and public controversy on a very grand scale, periodically visible to the public, mainly through official announcements, political debates and news media reporting, as the following quotation from Barbara Adam, Professor of Sociology at Cardiff University, illustrates:

> Bovine spongiform encephalopathy (BSE) in British cattle, commonly known as 'mad cow disease', is considered by many to have been the greatest disaster in the European food industry in the second half of the twentieth century. As such it presents an enormous challenge at every level of sociocultural organisation: farming, the food industry, science, politics and policy, public health, the media, and last – but by no means least – those at the receiving end, that is, consumers. Yet, despite the enormity of the disaster, BSE would not be on the public agenda had it not been for the media, which transformed this highly complex, intractable issue of scientific and political concern into public news.

(Adam, 2000, p. 117.)

From this quotation, it is clear that over a long period, communication, in particular through official announcements and political debates that were then reported in news media outlets including newspapers, television and radio, played an important role in informing the public about new developments regarding BSE and vCJD, influencing perceptions of this complex issue.

■ Can news media reports comprehensively represent the reality of complex scientific issues such as BSE/vCJD?

■ Given that newspaper articles are usually limited to a few hundred words and television and radio news bulletins to a few minutes, it would be impossible to comprehensively represent a complex issue such as BSE/vCJD in a news media report.

Media professionals place great emphasis on their ability to represent reality objectively. However, they also routinely make selections about which issues and events are worth reporting. Even when selected, as BSE/vCJD often was, media professionals then make further selections when reporting complex issues, not least because of the limitations of space in newspapers or time on television and radio. As a result, media reports are likely to represent issues in particular ways by placing greater emphasis on certain aspects over others. This is particularly important with complex emerging issues, such as BSE/vCJD, because members of the public would not have known that these diseases existed prior to media reporting. And research has suggested that mass media have greater influence on the audience when reporting issues where little would have been known previously. As a result, members of the public would have relied on media reporting of BSE/vCJD, which could only provide a partial view of the complex emerging issue.

Due to space & time the media are very selective in what they broadcast or print

In the following activity, you will investigate one of these reports, a newspaper article, which reported the March 1996 announcement to the House of Commons by the then Secretary of State for Health. In completing this activity, you will consider all four themes in relation to the science of BSE/vCJD.

Activity 3.1

Allow 45 minutes

Now return to the **BSE/vCJD Timeline** on the DVD-ROM, select the link to **Activity 3.1** and follow the on-screen instructions. Comments on this activity are provided with these on-screen instructions.

C R E D

You have investigated the newspaper article from Activity 3.1 in some detail, probably spending far more time than you usually do when reading a newspaper report. This information, alongside all that you have learnt from the *Introduction to the course* and Chapters 1 and 2, means that you have a good basic working knowledge of BSE/vCJD. What of the remainder of the public? What evidence is there to suggest that articles of this type influenced what people thought about BSE/vCJD?

3.2.1   Evidence of media influence

In Activity 3.1 you saw how the role of scientific experts, politicians and officials, media professionals and the food industry came to the forefront in reporting one aspect of the BSE/vCJD episode. In this respect, the views of the public, or consumers, have been downplayed, or their views have been taken as read.

■ Does the absence of credible information on what the public thought about BSE/vCJD seem like an oversight to you?

▪ Given that many members of the public could have been affected by BSE/vCJD, it would seem sensible to have canvassed their views on a range of related issues, for example, by investigating why so many people were deciding not to purchase beef-related products. Such information would have been invaluable in informing the campaign of reassurance from MAFF, the Chief Medical Officer and the MLC, for example.

- Can the news media be said to represent public opinion?

- Could you imagine a situation where you always agree with what you read in a newspaper or watch on the television news? Given the potential for the public to hold diverse views, it would be dangerous to assume that news media reporting is an accurate measure of public opinion.

If you think back to the *Introduction to the course* and the discussion of the theme of communication, you will recall that the producers of mass media (media professionals) are separated in time and place from the consumers (readers, listeners and viewers), although they may have channels for feedback, including email contact and letters. Taking mass media reporting as an indication of public opinion about BSE/vCJD, even when taking account of the different ways that media outlets reported this issue, is therefore likely to provide only a partial view of the likely diversity in public opinion.

How then would it be possible to investigate public opinion with regard to a complex and controversial issue such as BSE/vCJD? This raises the question of how assessments of what the public thinks about an issue can be investigated. The answer is that there is no one way to investigate public opinion. Below, you are introduced to two general approaches, each of which has its strengths and limitations. Indeed, researchers often seek to combine these approaches as a way of investigating different aspects of the same issue.

3.2.2 Large-scale survey approaches

Survey methods, which produce statistical (quantitative) data, provide one important approach to judging public attitudes to BSE/vCJD. In its most unsophisticated form, this could involve an analysis of the levels of beef consumption across the entire population. If levels of beef consumption drop following news media reporting of a public announcement, such as the first case of FSE, then it could be argued that the public had increased concerns about the safety of beef, and that media reporting was a factor in informing the public. However, these interpretations could only be inferred from these data.

A more sophisticated approach would involve systematic data collection with large samples, asking specific questions relevant to the issue being investigated. A key strength of this approach is that, by asking specific questions, a large-scale survey can produce results that are representative of the wider population, and this can then be compared to other data such as levels of beef consumption in the population as a whole. Examples of this approach include large-scale opinion polls that seek to identify whether individuals are scientifically literate or not, by asking questions such as 'what is a molecule?' and 'how long does it take for the Earth to go around the Sun?' Two such surveys have been carried out in the UK, in 1988 and again in 1996. Although not specifically designed to investigate media influence of BSE/vCJD reporting, the 1996 survey did show that interest in science and confidence in scientists remained high (when compared to the 1988 survey), but that there had been a loss of trust in scientists working for MAFF and the food industry. These findings suggest that the BSE/vCJD episode had had a profoundly negative effect on public opinion.

These survey methods can be limited, however, in terms of the types of questions that are asked by the researchers, which, in turn, can influence the answers that are given. For example, if researchers ask someone if they are willing to eat British beef, a survey can report on how many people are willing to eat British beef, but not necessarily on their reasons for taking such a decision, which could involve concerns about food safety, and/or ethical or religious reasons for not eating meat at all. It is also fair to say, then, that judging what the public thinks about an issue is very difficult given that individual opinions can involve sophisticated reasoning and can, at times, be contradictory. To investigate such views, researchers have adopted methods that allow research participants to discuss their knowledge, experience, attitudes and beliefs with respect to BSE and vCJD.

3.2.3   Small-scale qualitative approaches

In general, approaches that produce qualitative data do not attempt to be representative of the wider population. Nor do they tend to produce statistical (quantitative) data. Rather, they attempt to provide valid accounts of the complex views held by individuals through a range of methods, working with small, strategically chosen samples, producing mainly qualitative (e.g. written and spoken) data. These approaches tend to be less directed by the researchers in terms of the types of questions that are investigated. In this respect, the research participants are invited to express their views on a particular subject, using their own terms and vocabulary, for instance, within a group environment, through participant observation, or during individual interviews.

One such example, conducted by Jacquie Reilly from the Department of Sociology at Glasgow University, involved a series of focus group interviews with 258 participants. Initially, 26 focus group interviews involving 171 participants were conducted during 1992–1993 discussing issues related to food scares, such as salmonella in eggs and BSE. Thirteen of these groups (with 87 participants) were reconvened in 1996, following the 20 March 1996 announcement by the Secretary of State for Health. In this second series of focus groups, participants were asked to discuss whether their views on BSE/vCJD had changed since the earlier groups.

The results showed that media reporting was a key source of information about BSE. It also demonstrated that a number of participants had changed their views quite dramatically between 1992 and 1996. For example, one male participant argued in 1992 that:

> It's all a lot of nonsense really. I mean, to think that the government would allow us to be put at risk from such a thing is ridiculous. The media are just playing things up again and trying to scare us. I don't believe a word of it, and I certainly haven't stopped eating meat.

When interviewed again in 1996, the same participant argued that:

> I couldn't believe it when I saw it on the TV. It was such a shock to think that all these years I'd completely believed that there was no risk because I thought it was just a scare. I watched everything I could on the TV and realised that so much information was kept from us, that risks were taken with our health for purely economic reasons. I put everything with beef in it in the bin straight away and won't touch the stuff now…

(Both quotes cited in Reilly, 1999, p. 131.)

These two statements illustrate how media reporting of high profile announcements, such as that given by the Secretary of State for Health on 20 March 1996, can influence public attitudes, in this case in relation to BSE and vCJD. This example also shows that media reporting has the potential to influence behaviour; in this case, in terms of this person deciding to stop eating beef in 1996. Of course, it is important to remember that media coverage is only one influence amongst many. Friends, family members and other sources of information, including local butchers and farmers, also had the potential to influence people's views about BSE/vCJD. Furthermore, it is possible to identify a change in attitude with regard to official policies and the media. In 1992, this participant demonstrated trust in official policy and scepticism of media reporting; by 1996, television had become an important source of information about BSE. This loss of trust in authority is one of the most important legacies from the BSE/vCJD episode.

Of course, we have no way of knowing whether this person has started eating beef again since the time of the last interview. It is also important to bear in mind that this example only illustrates the reaction of one person, although the views of several other participants in this study were broadly along similar lines.

So far in Topic 1, we have considered the biology of prion diseases, the development of the BSE/vCJD episode and the short-term legacy of this issue: judging the long-term legacy of the BSE/vCJD episode is much harder. To address this aspect, we can refer to one key aspect of this debate that has yet to be discussed in any detail, the BSE Public Inquiry, and its relationship to science, public concerns and the making of public policy.

3.3 Science, public concerns and the making of public policy

As you have seen from Chapter 2, the making of public policy in relation to BSE/vCJD involved numerous revisions and updates, from scientists and politicians in particular, as new evidence emerged about these diseases. Consumption of beef also fell in the UK following the March 1996 announcement, illustrating public concern about the safety of beef.

■ Do you remember your own response to the March 1996 announcement? Did you stop eating beef or products derived from beef? Do you now eat beef or products derived from beef? What evidence did you use to inform these decisions?

▨ In choosing whether to eat beef or products derived from beef, you have made an important decision that affects you as an individual. Of course, this decision may also affect your family and friends; for instance, if you decide to cook beef for Sunday lunch. To make your decision, you may have been influenced by news media reporting, discussions with family and friends, even a chat with the local butcher. It is unlikely that these decisions were informed by detailed scientific evidence of prion diseases, however.

Despite the lack of detailed evidence regarding the public's opinions, concerns about safety have always been a crucial factor in the making of public policy in relation to BSE/vCJD. However, this was not the only factor. Alongside the emergence of new scientific knowledge, commercial considerations were also important, not least because the beef and dairy industry was a significant contributor to the UK economy. Policy makers therefore had to make decisions about BSE/vCJD in the light of incomplete scientific understanding and competing factors. This may help to explain the UK government's campaign of reassurance, which removed some of the scientific caveats in relation to the theoretical risks associated with TSEs.

something said as a warning or caution.

C R

■ Is this an example of a breakdown in communication between scientific advisers, policy makers and the public?

▨ Yes. In removing the caveats about the theoretical risks associated with TSEs, politicians made the assumption that the public would respond badly to scientists' descriptions of the potential risks of eating beef. As the evidence presented in Section 3.2 shows, this resulted in a loss of trust in scientists working for MAFF and the food industry.

Did this also lead to changes in public policies relating to scientific issues? Indeed, how can future policies be informed by high profile issues, such as the BSE/vCJD episode? One method for investigating these issues is a public inquiry. This form of retrospective analysis is important in establishing both successes and failures in managing uncertain issues, such as an emerging illness. Whilst holding a public inquiry can be enormously expensive, which is one of the reasons why few issues are subjected to this process, they have the potential to inform policy making in relation to future analogous events through the inquiry's recommendations.

3.3.1   The BSE Inquiry — *the Phillips Report (pub. Oct 2000)*

It was estimated in November 2000 that the BSE episode cost the beef industry £4 billion. This fact alone illustrates the enormous impact of BSE and may explain why in December 1997 an announcement was made in the UK Parliament that a public inquiry would be held to investigate issues relating to BSE/vCJD. The subsequent 16-volume final report, now known as the Phillips Report, was published in October 2000, documenting the committee's extensive findings. As with all public inquiries, terms of reference were agreed to guide the process of investigation. The terms of reference for the BSE Inquiry were:

> To establish and review the history of the emergence and identification of BSE and new variant CJD in the United Kingdom, and of the action taken in response to it up to 20 March 1996; to reach conclusions on the adequacy of the response, taking into account the state of knowledge at the time; and to report these matters to the Minister for Agriculture, Fisheries and Food, the Secretary of State for Health and the Secretaries of State for Scotland, Wales and Northern Ireland.

(Phillips Report, 2000, p. xiii.)

The terms of reference for any inquiry are crucial to the course of any subsequent investigation. They both guide the investigating team in the types of questions that they ask and influence the types of evidence sought in answering those questions. By the same token, they also rule out certain lines of enquiry and/or evidence.

Since the publication of the Phillips Report, MAFF has been restructured. The Department for Environment, Food and Rural Affairs (Defra), created in June 2001, currently has responsibility for, among other things, farming and food production.

In the case of the BSE Inquiry, the terms of reference were quite broad, allowing the Inquiry Committee scope to investigate both the development of scientific knowledge in relation to the emergence of BSE and vCJD, but also to consider the adequacy of the response, for example, by MAFF and the Department of Health (DoH).

■ Identify a factor restricting the Inquiry Committee's investigations.

▨ The Inquiry Committee was given licence only to investigate issues related to BSE and vCJD prior to 20 March 1996. As documented earlier in this topic, this was a key date in the development of BSE because it was the date that the then Secretary of State for Health, Stephen Dorrell, announced to the UK Parliament that a new variant of CJD had been identified and that further precautionary measures would be taken following advice from SEAC.

In choosing to limit the BSE Inquiry to events that occurred in relation to BSE/vCJD prior to 20 March 1996, the committee could focus on the key stages in the emergence of these two new diseases.

C R D

Having investigated 3000 files of documents and 1200 statements, and examined 333 witnesses over 138 days of evidence, the committee's findings and conclusions were extensive. They document, in particular, the ongoing risk assessments in relation to the spread of BSE within cattle and the possibility that BSE could be transferred to humans, issues that you encountered in Chapter 1. They also detail the precautionary measures taken by the UK government and European Union (Chapter 2), noting how these decisions were informed by expert scientific advice. In this respect, the report argues that:

R D

> There is a popular misconception that the Government did nothing to protect the public against the risk BSE might pose to human health until the likelihood of transmissibility was demonstrated in 1996. It is important to emphasise that the most significant measures to protect human health were taken at a time when the likelihood of transmissibility was considered to be remote.

(Phillips Report, 2000, p. 21.)

It could be argued, therefore, that the UK government had applied the precautionary principle to protect human health, for example by introducing the SBO ban, even though there was no clear scientific evidence to suggest BSE could be transmissible to humans, and that these early policy decisions were largely successful in reducing the risks to human health.

If the UK government was praised for taking precautionary measures to protect human health, why has the BSE/vCJD episode been described as 'the greatest disaster in the European food industry in the second half of the twentieth century' (Adam, 2000, p. 117)? In fact, the UK government was criticised for a number of reasons in the Phillips Report. Key among these was the government's policy of consistently reassuring the public that beef was 'safe to eat' and the perception that important information was being withheld from consumers, meaning that they were not given the chance to make decisions based on the best available advice at the time. In this sense, it could be argued that the public were exposed to an 'involuntary risk'.

The BSE Inquiry report goes on to argue that:

> The public was repeatedly reassured that it was safe to eat beef. Some statements failed to explain that the views expressed were subject to proper observance of the precautionary measures which had been introduced to protect human health against the possibility that BSE might be transmissible. These statements conveyed the message not merely that beef was safe but that BSE was not transmissible.

> The impression thus given to the public that BSE was not transmissible to humans was a significant factor leading to the public feeling of betrayal when it was announced on 20 March 1996 that BSE was likely to have been transmitted to people.

> (Phillips Report, 2000, p. xxi.)

Activity 3.2

Allow 10 minutes

Refer back to your notes from Activity 2.2, in which you answered questions on the synopsis 'Public reassurances about beef'. Compare these notes with your response to Activity 3.1, in which you examined the newspaper article, and also with the discussion of media influence (Section 3.2.1). Drawing on evidence from these activities, what effect do you think the campaign of reassurance had on consumer confidence about beef? (*Hint*: Consider both the impact on the beef industry if exports were affected and the levels of beef consumption in the UK.) Compare your answer with the comments on this activity at the end of this book.

As you have seen, the policy of reassurance was largely effective until the announcement in March 1996 when sales of beef dramatically reduced. In the months immediately following this announcement, public perceptions of the risks associated with eating beef altered and trust in politicians and scientists working for MAFF and the food industry was affected.

3.3.2   Changes in public policy

C R D

In reflecting on the lessons learnt from the BSE/vCJD episode up to March 1996, the BSE Inquiry called for changes in public policy. The report recommended greater *openness* and *transparency* in governing uncertain risks as a way of generating trust and credibility, recommendations that were based on the premise that the public would welcome a more open and honest debate about risk and uncertainty. In adopting this approach, discussions of absolute safety would shift towards descriptions of relative safety. These recommendations have had far-reaching implications for the relationship between science and society, affecting public policy in a number of areas. For example, in 2000, the House of Lords Select Committee on Science and Technology produced an influential report called 'Science and society'. This report highlighted the 'crisis of confidence' that existed between science and society. It argued that many of these concerns were seen, by the public, to be the result of a perceived lack of transparency in the relationship between science, industry, public policy and the public as consumers. High-profile issues, such as the BSE/vCJD episode, were also seen as responsible for reducing levels of trust in scientific expertise. It was argued that effective science communication could be a key factor in reducing tensions between science and society. This has influenced how issues of food science are governed, for example, with the introduction of the Food Standards Agency (FSA), which is committed to promoting openness and transparency in providing independent advice on a range of food issues, including BSE/vCJD. It also has a commitment to consult the public on issues of food health, labelling, etc. This can be seen as a commitment to increasing public dialogue and engagement with scientific issues, another important legacy of the BSE/vCJD episode and an issue that you will revisit later in the course.

Summary of Chapter 3

1   This chapter has examined the legacy of the BSE/vCJD episode by examining three aspects of this issue. First, you were introduced to developments in the scientific community, where changes have been made in the ways that scientific knowledge informs policy. The introduction of SEAC, the CJD Surveillance Unit, the FSA and the restructuring of MAFF to become Defra are important examples of these changes. Funding has also increased for investigations into TSEs and prions.

2   Second, quantitative and qualitative evidence indicated a shift in attitudes in the relationship between science and society characterised by a loss of trust in politicians, public officials and scientists working for MAFF and the food industry, changes that may have long-term consequences.

3   Third, you have addressed issues of science, public concern and the making of policy, noting how lessons learnt from the BSE Inquiry have informed recent policy making in relation to emerging scientific issues.

4   The House of Lords report 'Science and Society' indicates a shift in public policy where openness and transparency, and dialogue and consultation, are highly valued, linking the themes of communication, risk, ethical issues and decision making. You will revisit these issues of as you continue your study of this course.

Questions for Chapter 3

Question 3.1

Briefly outline three changes affecting the scientific community that are related to the BSE/vCJD episode.

Question 3.2

(a) Name two general approaches that can inform our understanding of people's perceptions of complex science-based issues, such as BSE/vCJD? (b) What types of evidence can they generate?

Question 3.3

The 20 March 1996 announcement had a profound effect on the beef industry and public perceptions. Briefly outline two examples that illustrate these changes.

Question 3.4

What long-term effects has the BSE/vCJD episode had on public policy in relation to science and society?

Learning Outcomes for Topic 1

S250's Learning Outcomes are listed in the *Course Guide* under three categories: Knowledge and understanding (Kn1–Kn6), Cognitive skills (C1–C5) and Key skills (Ky1–Ky6). Here, we outline how these overall learning outcomes have been treated in the context of Topic 1 BSE/vCJD and its associated assessment.

The way that prion molecules cause diseases such as BSE and vCJD, and how the key discoveries about prions were made, relate primarily to Kn1 and Kn2. The patterns of BSE and vCJD in populations, and how this information is used to predict the number of cases there may be in future (and to assess the accuracy and precision of such predictions), also relate to Kn2. Whilst science can make important contributions to managing episodes such as BSE/vCJD, so can disciplines outside the natural sciences and also members of the general public, including farmers and consumers of beef-related products (Kn6). This broad societal context is represented largely by the course themes of communication (Kn3, Kn4), risk (Kn4), ethical issues (Kn5) and decision making (Kn3, Kn6). These themes were introduced in the *Introduction to the course*, exemplified by the BSE/vCJD episode, and will be further developed in later topics.

Having studied Topic 1, you should have developed your abilities to evaluate, interpret, synthesise – and also recognise deficiencies in – information and data (C1); assess the extents to which claims and arguments are based on scientific evidence (C2); make defensible judgements based on scientific and other information (C3); and explain the contribution that science can make to addressing major contemporary issues such as BSE/vCJD and also its limitations (C4). You should now be able to apply the knowledge and understanding you have gained about BSE/vCJD to future developments in prion biology and also to unfamiliar topics, provided you are given additional information (C5).

Although, while studying Topic 1, you were expected to receive, respond to, select and use relevant information presented mainly in printed course material, for assessment purposes information may come from other sources and be provided through other media (Ky1). You must be able to interpret and possibly present data that may be qualitative and/or quantitative (Ky2), and also to communicate clearly, correctly and logically to specified audiences (Ky4). Although there has been less emphasis on mathematical skills in Topic 1 than there is in later topics, you do have to be able to read and possibly produce simple graphs and tables, perform straightforward manipulation of data (e.g. calculate percentages and ratios) and understand basic statistical concepts (such as means and 95% confidence limits) (Ky3). At this early stage in the course, there is likely to be less scope than there will be later to work with others (Ky5). As always, your aim should be to use every opportunity to plan, monitor and develop strategies for more effective learning (Ky6).

Answers to questions

Question (i)

(a) The process whereby new scientific knowledge is verified prior to publication is called peer review. (b) Peer review involves anonymous expert reviewers who check work submitted for publication in the form of a manuscript for quality and mistakes. If the work is of insufficient quality it is normally rejected. If mistakes are found, then the reviewers request revisions by the original author. If the manuscript 'passes' peer review, it is normally accepted for publication.

Question (ii)

(a) Near-Earth objects (NEOs) provide an example of a naturally occurring risk. (b) The probability of a large NEO colliding with the Earth in the next 10 years is extremely small.

Question (iii)

(a) The three guiding principles that the researcher should use are: reduction, refinement and replacement. (b) Refinement requires that the researcher considers how they could minimise suffering and distress to the animals, whereas reduction involves using the fewest animals possible to complete the research effectively. Alternatively, the researcher could try to find a replacement means to conduct their research, and not use animals at all.

Question (iv)

(a) The precautionary principle can be used to inform decision making about complex science-based issues. (b) The precautionary principle requires that, in the absence of full scientific evidence, a precautionary approach to decision making about risk should be adopted.

Question 1.1

(a) (i) Percentage of BSE cases that involved female cattle:

$$\frac{190}{192} \times 100\% = 98.9583\%$$

Since both starting values are three-digit numbers that are known precisely, the answer should be expressed to 3 significant figures: 99.0%.

(ii) The total number of cattle in the national herd was
$3\,200\,000 + 37\,000 = 3\,237\,000$.

Percentage of the national herd that was female:

$$\frac{3\,200\,000}{3\,237\,000} \times 100\% = 98.8570\%$$

Since both starting values are given to at least 2 significant figures (some of the trailing zeroes *might* be significant), the answer can safely be expressed to no more than 2 significant figures: 99%.

(*Note*: Any ambiguity about the number of significant figures to which some of the starting values are expressed could be removed by using scientific notation. Thus, expressing 3 200 000 as 3.2×10^6 rather than 3.20×10^6 would make clear that the number of significant figures was 2 rather than 3. However, the use of scientific notation in this way is not the norm in all fields of study.)

(iii) Since the percentage of female cattle in the sample of 192 BSE cases is almost the same as the percentage of female cattle in the national herd, there is no evidence in these data that one sex was more prone to BSE than the other.

(b) (i) Proportion of BSE cases in beef suckler cows:

$$\frac{14}{880000} = 0.000016$$

Since one of the starting values is given to 2 significant figures and the other to at least 2 significant figures, the answer is appropriately expressed to 2 significant figures.

Note: Since a proportion is defined as the size, number or amount of one object or group as compared to the size, number or amount of another, it can legitimately be expressed in a variety of ways – as a decimal fraction (0.000 016), as a decimal fraction given in scientific notation (1.6×10^{-5}), as a percentage (0.0016%), as a percentage given in scientific notation (1.6×10^{-3}%), as a ratio (14 : 880 000, which simplifies to 1 : 63 000) or as a conventional fraction, i.e.

$$\frac{14}{880000}, \text{ which is equivalent to both } \frac{16}{1000000} \text{ and } \frac{1}{63000}.$$

(ii) Proportion of BSE cases in dairy cows:

$$\frac{696}{2320000} = 0.000300$$

As one of the starting values is given to 3 significant figures and the other to at least 3 significant figures, the answer is appropriately expressed to 3 significant figures.

(iii) The proportion of BSE cases in dairy cows to the proportion of BSE cases in beef suckler cows = 0.000 300 : 0.000 016 or 19 : 1 (appropriately expressed to 2 significant figures).

(c) Since dairy calves are removed from their mothers soon after birth, they have to be fed concentrates. In contrast, beef calves are seldom fed concentrates. The *much* higher incidence of BSE in dairy cattle compared to beef suckler cattle is therefore consistent with the cause being contaminated MBM incorporated into cattle concentrates. Given the relatively long incubation periods of TSEs, it may also be relevant that many dairy cattle were allowed to live far longer than beef suckler cattle.

Question 1.2

(a) The main error in the statement is that there are no such thing as alleles of the *PrP* gene known as *PrP*^{C} and *PrP*^{Sc} and therefore one cannot mutate into the other. Through transcription and translation, many genes produce proteins. Most genes exist in different versions or alleles. Mutation can cause one allele to change into another allele. In turn, this can cause a protein to be produced that has a slightly different sequence of amino acids. The *PrP* gene codes for the PrP protein. Different alleles of this gene produce different versions of the PrP protein which differ in their sequence of amino acids.

However, PrP^C and PrP^{Sc} refer to different *conformations* that a PrP protein can adopt *irrespective of its amino acid sequence*. PrP^C is the 'normal' conformation of the PrP protein and PrP^{Sc} is the conformation that results in prion diseases, such as BSE and vCJD. A PrP^C molecule can adopt the PrP^{Sc} conformation either spontaneously or as a result of interaction with another molecule already in the PrP^{Sc} conformation.

(b) A correct version of the statement would be: 'Prion diseases such as BSE and vCJD are caused when the 'normal' version of the PrP protein, coded for by the *PrP* gene and known as PrP^C, takes on an alternative conformation, when it is known as PrP^{Sc}'.

Question 1.3

(a) The PrP^{Sc} molecule that starts the 'chain reaction' could have been produced *within* the cell by a 'normal' PrP^C molecule spontaneously changing into the PrP^{Sc} conformation. Alternatively, it could have arrived in the cell from *elsewhere*. In the latter case, the PrP^{Sc} molecule could have been released by another cell in the same animal or it could have been produced by another animal and transferred from animal to animal in food.

(b) Further PrP^{Sc} molecules are produced through interaction between individual PrP^{Sc} molecules and PrP^C molecules synthesised within the cell. This interaction somehow causes the PrP^C molecules to adopt the PrP^{Sc} conformation.

(c) PrP^{Sc} molecules have a tendency to collect together as insoluble deposits which eventually kill brain cells. PrP^{Sc} molecules taken up by other cells can cause the 'chain reaction' to start in these cells too. The symptoms of prion diseases become apparent when large numbers of brain cells have been killed in this way.

Question 1.4

(a) The 2003 Imperial College prediction was as follows: best estimate, 200; upper and lower 95% confidence limits, 7000 and 10 respectively.

(b) Even 3000 vCJD deaths by 2080 are fewer than the upper 95% confidence limit. Nevertheless, this number of deaths is much higher than the best estimate of 200 deaths. So, although 3000 vCJD deaths does not invalidate the prediction, such a high number does call into question the assumptions upon which it was based.

(c) Even though the best estimate was 200, there could have been 3000 deaths *by chance* without undermining the assumptions. However, it could well be that one or more of the assumptions built into the team's model were wrong. An important assumption was that all the victims became infected by eating contaminated meat. It is already possible that some may have acquired their vCJD through blood transfusion. Another important assumption was that only people homozygous at triplet 129 of the *PrP* gene for methionine (*MM*) were susceptible to vCJD. It is already known that at least one person heterozygous at this triplet (*MV*) acquired vCJD. People homozygous for valine (*VV*) may also be susceptible.

Question 2.1

(a) The ruminant feed ban, introduced in July 1988, banned the inclusion in feed for ruminant animals (e.g. cattle, sheep and goats) of proteins (other than in milk) derived from ruminant animals.

(b) It was believed that BSE was transmitted as a consequence of the incorporation in cattle concentrates of MBM contaminated with some kind of biological agent (later identified as PrP^{Sc} protein) that had not been inactivated by the recently altered rendering process. Banning the inclusion of proteins from all ruminant animals in the feed given to other ruminant animals should have stopped the cycle of BSE transmission.

Question 2.2

(a) The number of new BSE cases reported annually grew until 1992 and then continued at a high level for several years after that (Figure 1.1). Although the relatively long incubation period of BSE makes interpretation of these data somewhat difficult, cases of BSE in cattle born after the introduction of the ban – particularly after controls were tightened in 1996 (Section 2.3) – provide incontrovertible evidence that either the ban was not working effectively or that cattle feed was not the only way in which animals could contract BSE.

(b) To some extent the ban was ignored on farms, for instance to use up old feed. There must also have been accidental contamination – both on farms and during manufacture – of feed intended for ruminants with feed intended for other animals (e.g. chickens) from which ruminant-derived materials had not been banned. Some SBO-containing material was exported as fertiliser and then re-imported as feed.

Question 2.3

Application of the precautionary principle involves going beyond available knowledge of risk so as to err on the side of caution. SBO derived from cattle with BSE was believed to be a relatively rich source of PrP^{Sc} protein (though not as rich as the brain and spinal cord), and had therefore been banned from food for human consumption. However, although SBO from scrapie-infected sheep has been proved to be infective, only nervous tissue has been found to be infective in the case of BSE (Section 1.4). So banning cattle SBO in the first place might be regarded as an application of the precautionary principle. Having decided on the SBO ban, staining it blue can be seen as an enforcement measure.

Question 2.4

(a) None. A few treatments had been tried that were claimed to slow down – or even stop – progress of vCJD in some patients. These treatments were based on medicines developed, tested and approved as treatments for other conditions (e.g. quinacrine for malaria and PPS for urinary tract infections).

(b) If it were known that many thousands of people are likely to develop vCJD over the coming decades, then the huge investment needed to develop a cure for the disease might be justified. However, if only about 200 people are likely to develop vCJD – with about three-quarters of these having died already – then most people would agree that there are higher medical priorities for this investment. Trying possible treatments for vCJD on an *ad hoc* basis on small numbers of terminally ill patients also raises issues related to informed consent and to the objective evaluation of treatments.

Question 3.1

Scientists now know considerably more about prion diseases as a result of the BSE/vCJD episode. The 'scientific infrastructure' required to investigate these diseases is also now in evidence and continues to develop. Furthermore, new bodies such as SEAC and the CJD Surveillance Unit now exist to provide advice and guidance to policy makers and collate new research findings.

Question 3.2

(a) Large-scale surveys and small-scale qualitative approaches can inform our understanding of people's perceptions of complex science-based issues such as BSE/vCJD.

(b) These general approaches generate quantitative (statistical) and qualitative (e.g. spoken or written) data.

Question 3.3

Beef sales fell dramatically following the 20 March 1996 announcement, profoundly affecting the beef industry. Public perceptions, in particular in relation to scientists working for MAFF and the beef industry, also changed, indicated by a loss of trust in these experts.

Question 3.4

Public policy in relation to science and society now routinely involves openness and transparency in communicating complex issues to the public as a way of promoting informed consent. A commitment to increasing dialogue and consultation now also forms an important aspect of public policy, involving members of the public in decision making about science-related issues.

Comments on activities

Activity 1.1 (Part 2)

As noted in the text, although cattle displaying symptoms of BSE are usually killed immediately (p. 32), this is not a policy applied to humans who contract fatal diseases.

Ethical issues may arise from medical use of biological materials (such as growth hormone or corneas) derived from the bodies of deceased people (presumably with their explicit consent) *even if their undiagnosed CJD had not caused iatrogenic CJD in the patients* (p. 36).

Cannibalism is one of the great taboos in most cultures and societies. Routine cannibalism among the Foré until the mid-20th century (pp. 36, 37) would therefore certainly be regarded an ethical issue in a pejorative sense by many UK citizens. On the other hand, the Foré presumably engaged in this activity as a mark of great respect for their dead, and from their point of view not to engage in cannibalism would be an ethical issue. This illustrates how value systems from different cultures and societies influence how ethical issues are defined.

As discussed in Box 1.2 (p. 42) and elsewhere, the deliberate infection of animal 'models' with disease – and, indeed, the use of non-human animals in experiments generally – is an ethical issue because of the suffering and possible deaths of these animals. Whilst this is manifestly so for people who object on principle to all or most such experimentation, this should also be true for all responsible citizens and especially for those engaged in designing or carrying out experiments involving animals.

Activity 1.1 (Part 3)

Since cattle are herbivorous, some people would object to them being fed tissues derived from other animals (p. 47) because it challenges their normal behaviour. Indeed, most people were probably unaware of this practice until BSE started to be discussed. Examples of experimental procedures that some people would consider ethically unacceptable include the injecting of brain tissue from an infected animal into the brain of another in order to investigate the incubation period of TSEs in relation to lifespans (pp. 53–4) and injecting infective prions through a succession of species to investigate perpetuation of strains (p. 66). The issue of cannibalism and whether it should be regarded as taboo or a mark of respect for the dead comes up again (p. 58).

A very important ethical issue is whether or not information that someone was incubating vCJD should be provided to them or withheld (p. 64). At present, it is not possible to cure people of vCJD and there is only very limited evidence that progress of the disease can even be slowed in those already displaying its symptoms (see Chapter 2). Given the profound psychological damage this knowledge might cause, the distinct possibility that the person might in any case die from an unrelated cause before developing vCJD and the fact that precautions have been put in place to minimise their infecting someone else in the meantime, there is a real challenge in deciding whether to pass on this information. Of course, the ethical balance could change dramatically if a treatment became available that could prevent the development of vCJD but *only* if it were given before the appearance of symptoms.

A further consideration would be the cost of any treatment and whether it should be available to all or only to those who could afford it. Indeed, a very important issue is how much resource – that could be used for other purposes, medical or otherwise – should be devoted to trying to develop either a preventative treatment or a cure for a disease such as vCJD. And who should foot the bill? As noted in the text (p. 65), this does rather depend on the eventual total number of people likely to develop the disease.

Activity 1.2

If the reactions of other scientists really did include 'outrage' at the *ideas* presented in a peer-reviewed scientific paper (rather than to, say, any courting of the media that Prusiner might have engaged in), then this suggests the paper challenged really deeply held ideas. The same could be said of the article in *Discover* magazine, even though its title suggests it was primarily concerned about the amount of publicity given to a concept supported by limited experimental evidence and to the concept's originator. The concept itself – that TSEs are caused by an infectious agent which contained no genetic material – was certainly controversial. Prusiner's self-imposed ban on talking to the media suggests that he might have been hurt by this criticism – but not so severely as to stop working in the area. Indeed, he was presumably convinced that he was right. We have to assume Gajdusek's unwillingness to accept Prusiner's 'protein-only' hypothesis is attributable to genuine scientific scepticism. Of course, the eventual award of a Nobel Prize for his work gave Prusiner tremendous prestige both within and beyond the scientific community. Among the many scientists who now accept Prusiner's explanation of TSEs, and readily use his newly coined word 'prion', are presumably some of those who were originally 'outraged' by his ideas.

Activity 2.1

The comments for this activity are included as part of the **BSE/vCJD Timeline**, which you will find on DVD-ROM.

Activity 2.2

The comments for this activity are included as part of the **BSE/vCJD Timeline**, which you will find on DVD-ROM.

Activity 2.3

The comments for this activity are included as part of the **BSE/vCJD Timeline**, which you will find on DVD-ROM.

Activity 3.1

The comments for this activity are included as part of the **BSE/vCJD Timeline**, which you will find on DVD-ROM.

Activity 3.2

In the first instance, the campaign of reassurance could be deemed to have been relatively successful in terms of beef consumption in the UK because beef sales remained relatively unaffected. In effect, it could be argued that the public perceived eating beef to be a relatively safe activity. This is also borne out in the research conducted by Jacquie Reilly. As the male participant in the focus group study argued in 1992, 'I certainly haven't stopped eating beef'.

It could be argued that the campaign of reassurance in May 1990 was partly influenced by the introduction of export restrictions by the EC, for instance in March and April 1990. These restrictions would have had enormous significance for the beef industry in terms of its overall sales and the UK government would have been keen to take measures to improve consumer confidence. However, even with the ongoing campaign of reassurance, the EC continued to restrict exports of beef and these actions would have influenced perceptions of the risks associated with British beef, particularly within the EU.

However, it was not until the March 1996 announcement that BSE could be the cause of vCJD, and subsequent media reporting of this event, that consumption of beef in the UK was significantly affected. As the same male respondent from the focus group study argued in 1996, 'I put everything with beef in it in the bin straight away and won't touch the stuff now'.

References and further reading

Adam, B. (2000) The media timescapes of BSE news, in Allan, S., Adam, B., and Carter, C. (eds) *Environmental Risks and the Media*, Routledge, London and New York.

Anderson, R. M., Donnelly, C. A., Ferguson, N. M. *et al.* (1996) Transmission dynamics and epidemiology of BSE in British cattle, *Nature*, **382**, pp. 779–788.

Boyne, R. (2003) *Risk*, Open University Press, Buckingham.

Fuller, S. (1997) *Science*, Open University Press, Buckingham.

House of Lords Select Committee on Science and Technology (2000, 23 February) *Science and Society* (3rd Report), HMSO, London.

Irwin A. and Wynne, B. (eds) (1996) *Misunderstanding Science*, Cambridge University Press.

Nelkin, D. (1995) *Selling Science: How the press covers science and technology* (rev. edn), W. H. Freeman, New York.

Phillips Report (2000, October) *The BSE Inquiry*, HMSO, London.

Prusiner, S. B. (1982) Novel proteinaceous infectious particles cause scrapie, *Science*, **216**, pp. 136–144.

Reilly, J. (1999) 'Just another food scare?' Public understanding and the BSE crisis, in Philo, G. (ed.) *Message received – Glasgow Media Group research 1993–1998*, Longman, Harlow.

Ridley, R. M. and Baker, H. F. (1998) *Fatal Protein: The story of CJD, BSE and other prion diseases*, Oxford University Press.

Warnock, M. (2001) *An Intelligent Person's Guide to Ethics*, Duckworth Media Group, London.

Acknowledgements

Dr Rosalind M. Ridley and her colleague Harry F. Baker provided invaluable advice and guidance to the authors. Dr Tom Chamberlain (University of Reading) and Dr Anne Martis (Associate Lecturer, Open University) were authors of the first and second editions respectively of the equivalent material in a previous Open University course, S280 *Science Matters*.

Grateful acknowledgement is made to the following sources for permission to reproduce material within this book:

Figure 1.4 Courtesy of Dr J. W. Ironside, National CJD Surveillance Unit; *Figure 2.1* Empics.com; *Figure 2.2* Courtesy of Foods Standards Agency, © Crown Copyright.

Every effort has been made to contact copyright holders. If any have been inadvertently overlooked, the publishers will be pleased to make the necessary arrangements at the first opportunity.

Abbreviations

A	adenine (nucleotide)
BSE	bovine spongiform encephalopathy
C	cytosine (nucleotide)
CJD	Creutzfeldt–Jakob disease
CMO	Chief Medical Officer
CVO	Chief Veterinary Officer
CWD	chronic wasting disease
DBES	Date-Based Export Scheme
Defra	Department for Environment, Food and Rural Affairs
DNA	deoxyribonucleic acid
EC	European Commission
EEC	European Economic Community
EU	European Union
FFI	fatal familial insomnia
FSA	Food Standards Agency
FSE	feline spongiform encephalopathy
G	guanine (nucleotide)
GSS	Gerstmann–Sträussler–Scheinker syndrome
Hb^A	'normal' allele of human haemoglobin gene
Hb^S	sickle-cell allele of human haemoglobin gene
MAFF	Ministry of Agriculture, Fisheries and Food
MBM	meat and bone meal
MLC	Meat and Livestock Commission
MM	homozygous for methionine at triplet 129 of *PrP* gene
MRM	mechanically recovered meat
mRNA	messenger ribonucleic acid
MV	heterozygous for methionine and valine at triplet 129 of *PrP* gene
NGO	non-Governmental organisation
PrP	gene coding for protease-resistant protein
PrP	protease-resistant protein
PrP^C	cellular (or 'normal') protease-resistant protein
PrP^{Sc}	scrapie-causing protease-resistant protein
SBM	specified bovine material
SBO	specified bovine offals
SEAC	Spongiform Encephalopathy Advisory Committee
T	thymine (nucleotide)
TME	transmissible mink encephalopathy
tRNA	transfer ribonucleic acid
TSE	transmissible spongiform encephalopathy
vCJD	variant Creutzfeldt–Jakob disease
VV	homozygous for valine at triplet 129 of *PrP* gene

Index

Entries in **bold** are key terms defined, along with other important terms, in the Glossary. Page numbers referring only to figures and tables are printed in *italics*.